Reporting

A Biologist's Guide to Articles, Talks, and Posters

Want to learn how to present your research successfully? This practical guide for students and postdocs offers a unique step-by-step approach to help you avoid the worst, yet most common, mistakes in biology communication. Covering irritants such as sins of ambiguity, circumlocution, inconsistency, vagueness and verbosity, misuse of words and quantitative matters, it also provides guidance to design your next piece of work effectively. Learn how to write scientific articles and get them published, prepare posters and talks that will capture your audience and develop a critical attitude towards your own work as well as that of your colleagues. With numerous practical examples, comparisons among disciplines, valuable tips and real-life anecdotes, this must-read guide will be a valuable resource for both new graduate students and their supervisors.

R. S. Clymo is Professor Emeritus at, and Honorary Fellow of, Queen Mary University of London. His distinguished career spans four decades and includes experience as researcher, teacher, reviewer, and editor of scientific journals. He has authored or co-authored numerous papers and journal articles, as well as a manual of water analysis. He is an Honorary Member of the British Ecological Society, Honorary Doctor of the University of Lund, and Bronze Medallist of the University of Helsinki.

Reporting

Research

A Biologist's Guide to Articles, Talks, and Posters

R. S. Clymo
Queen Mary University of London, UK

CAMBRIDGE UNIVERSITY PRESS

CAMBRIDGE
UNIVERSITY PRESS

University Printing House, Cambridge CB2 8BS, United Kingdom

Cambridge University Press is part of the University of Cambridge.

It furthers the University's mission by disseminating knowledge in the pursuit of education, learning and research at the highest international levels of excellence.

www.cambridge.org
Information on this title: www.cambridge.org/9781107053892

© Cambridge University Press 2014

This publication is in copyright. Subject to statutory exception
and to the provisions of relevant collective licensing agreements,
no reproduction of any part may take place without the written
permission of Cambridge University Press.

First published 2014

Printed in the United Kingdom by TJ International Ltd. Padstow Cornwall

A catalogue record for this publication is available from the British Library

Library of Congress Cataloguing in Publication data
Clymo, R. S.
Reporting research : a biologist's guide to articles, talks and posters / R. S. Clymo.
　pages　cm
Includes bibliographical references and index.
ISBN 978-1-107-05389-2 (Hardback) – ISBN 978-1-107-64046-7 (Paperback)
1. Biology–Authorship.　2. Technical writing–Vocational guidance.　I. Title.
QH304.C685 2014
570.1–dc23　2014006667

ISBN 978-1-107-05389-2 Hardback
ISBN 978-1-107-64046-7 Paperback

Additional resources for this publication at www.cambridge.org/9781107053892

Cambridge University Press has no responsibility for the persistence or accuracy of URLs for external or third-party internet websites referred to in this publication, and does not guarantee that any content on such websites is, or will remain, accurate or appropriate.

Contents

Preface *page* xi

PART I Basics

1. Writing a scientific article and getting it published 3
  The standard scientific article 4
    Title 7
    Summary or abstract or (rarely) synopsis, and keywords 10
    Symbols and abbreviations 12
    Introduction 12
    Materials, study area, and methods 16
    Results 18
    Discussion 21
    Conclusions 23
    Acknowledg(e)ments 23
    References (or literature cited) 24
    Tables and figures 25
    Supplementary materials 26
  Latin names of organisms 26
  What journal should I choose, and how many articles should I write? 28
  How long does it take to write and publish an article? 37

CONTENTS

Some advice 39
    After choosing a target journal, and up to the first public draft 39
    Revision of the first public draft 44
    Submission to your target journal 45
The editor and the publishing process 47
The editor's duties 53
Proof correcting 54
Intellectual property: copyright and patents 58
The digital electronic future 62

2. Speaking about your work 69
    Writing a summary 70
    Preparing what to say 70
    Visual aids 73
        Slide order and design 74
    Practising and refining 80
    On the day but before your talk 81
    The performance itself 82

3. Making and displaying a scientific poster 87
    The title and summary 88
    Preparing the poster content 89
    Designing and laying out the poster 92
    Before leaving for the meeting 97
    Setting up and tending the poster 98

4. Scientific authorship 101
    How many authors does it take to write an article? 101
    Who qualifies as an author? 107

PART II Improving

5. Style in writing 119
    The arrangement of ideas and words 124
    'Proper words in proper places' 125
        Guidelines 128

Punctuation and typographic details: the journal's rules 135
   Punctuation 136
   Punctuation that affects structure: comma, semicolon, colon, and stop 137
   The apostrophe 139
   Hyphens and dashes 140
   Enclosures 142
   Footnotes, endnotes, and other jumps 143
   Prefixes to people's names 144
   Abbreviations and italicised words 145
   Dates and clock time 146
   Typefaces and fonts 147
   Special characters 152

6. Frequently misused words and technical terms 157
   Frequently misused words 158
   Frequently misunderstood technical terms 177

7. Quantitative matters 189
   Numerical values 189
   Symbols 192
   Units and multipliers 193
   Dimensions and equations 200
      Dimensions must balance 200
      Combining units and dimensions 201
      'Equations' that are not: when dimensions do not balance 202
      Using dimensions to decide what to do: an example 205
   Expressing the name and value of a variable 206
   More about dimensions 208
      Dimensions of the coefficients in a polynomial 208
      Dimensions other than [M], [L], [T] 209
      Inferring the dimensions of a parameter 210

8. Managing error 211
   The sorts of error 212
      Mistakes 213

Imprecision and bias   214
Measurement error, material variation and error, and randomisation   218
Measures of dispersion and imprecision   221
Dispersion and describing: standard deviation (SD) and quantiles   221
(Im)precision and testing: standard error of the mean (SEM)   226
Significant digits   230
Combining errors   237
Examples   240

9. **Data interrelations**   243
Interpreting *P* values   243
Are two means significantly different?   246
Correlation, regression, and functional analysis   250
Correlation   250
Regression   251
Functional relations and analysis   253
A better way?   257
'Significance' and 'importance'   260
'Common sense'   260

10. **Tables and figures: the evidence**   261
Tables   263
Table heading   265
Table layout   266
Table furniture   270
Data items in a table   271
Figures: graphs   273
Design: general   275
Arrangement of multiple graphs   278
Design elements   278
Types of graph   290
Figures: other line diagrams   294
Figures: images   294
What image manipulations are acceptable?   294

CONTENTS

Appendix to Chapter 10: colour modes, resolution, and file formats 297
Colour modes 297
File formats and image resolution 300

11. Citing and referencing 305
Citations 306
Author–date system 306
Citation-number and reference-number system 307
References 308
Abbreviated titles of journals 312

ENVOI 315
Bibliography and references 317
Bibliography 317
References 318
Index 327

Preface

This book grew out of a repeated need, as graduate supervisor, referee (reviewer), and editor, to comment about irritants such as:

- the sins of ambiguity, circumlocution, confusion, inconsistency, vagueness and verbosity;
- misuse (or distractingly poor use) of words such as content, decimate, impact, level, light intensity, paradigm, parameter, and ratio;
- quantitative matters such as equations that didn't equate, ludicrous precision of numerical values, misleading bar charts, graphs with inaccurate axis labels, nonsensical 'log scales', and undefined error bars;
- statistics-related misuses such as the difference between standard deviation and standard error, statistical fishing expeditions, the meaning of a 'P-value', and confusion among correlation, regression, and functional relations;
- incomprehensible talks and unattractive, or even repellent, posters.

What follows includes my gleanings during four decades as a university biologist: the things I wish now that I had known when I started out. It is mostly about broadcasting for the first time the results of original work in the biosciences (the inelegant umbrella term 'bioscience' embraces all the disciplines that work on living things, including biochemistry, molecular biology, genetics, microbiology, biology, botany, zoology, anatomy, morphology, physiology, ecology, and applications in, for example, medicine, psychology, and soil science). This book is intended mainly for those beginning a career in bioscience research. If you are such an apprentice or journeyman then I hope that you learn from the mistakes I made at your age. Established bioscientists may also find things to interest them.

Some of what follows is elementary and you will know it already (but your colleagues may not). You will also find less familiar matter, some of it difficult to locate and some original. Knowledge about topics such as significant digits and the combination of errors is rare among bioscientists. The text contains sufficient detail for you to try to understand *why* these things are as they are. I also include more about the basis of simple statistics than you might expect because this is an area frequently criticised by editors and referees. Some of this will be of more help in designing your next bit of work than in reporting that already done. You will not find anything about grant applications, or how to seek approval from an ethics committee, or literature searching, or making a risk assessment, or journal clubs, or networking, or the need to keep wide interests, or any of the other activities that may contribute to a career in science.

Many of the examples come from the ecological end of the molecular-to-ecological range of the biosciences because that is where my research has been, but most of the principles they illustrate apply across the whole of the biological (and often other) sciences. Some examples are from fields outside bioscience in the hope that you may more easily concentrate on the principle they illustrate rather than the content. Yet others I have invented or modified from the original source.

In many places I have used a didactic (instructional) – or even imperative (commanding) – tone to save space and your time, and have assumed a 'typical' journal or scientific meeting. Yet many of the topics are more complex than I describe, or are contentious, or differ from subject to subject (biochemistry to ecology), journal to journal, or meeting to meeting. Science publishing is in ferment, so some things may be out of date or irrelevant in your particular field, and many of the views are personal. You may well disagree with them. Good, for that is the high road to enlightenment. Presenting bioscience research is an art and, as with any art, you improve with practice, thought, and discussion. I hope to stimulate in you a critical attitude to the writing, speaking, and posters of other people as well as to your own.

A didactic approach may easily result in lists and a cookbook: '(1) Do this, then (2) do that.' This is the style of most instrument manuals and many books about scientific writing. My experience has been that one uses an instrument more effectively if one understands how it works, and why things should be done, and which of them must be done in order or with all possible care, and which are less critical. So in this book I have included explanations, discussion, and a good deal of background and parenthetic matter in the belief that the more you know and understand the better your own reporting will be.

I experiment with three unusual features. The first concerns parenthetic ▼ material.

> ▼ Footnotes are a nuisance; endnotes are worse (Chapter 5). I therefore use indented panels like this one to contain parenthetic material that is cued by superscript geometric shapes ▼, ●, ■, and so on in the *immediately preceding* paragraph. Solid geometric shapes are easier to return to than letters or special typographic symbols such as '§'.

Second, where a figure contains more than one independent graph the conventional order is from the top left as if reading text. But in this book the graphs are usually ordered from the lower left, across, then upper left and across. This puts the first part of the explanatory caption (below the figure) nearest to the graph that it concerns, and orders whole figures in the same way that one reads a single graph (start at the lower left corner).

And third, to shift your focus occasionally from reading to reflecting I have included a few questions in boxes (this one is borrowed from Lewis Carroll's 'The Walrus and the Carpenter'):

QUESTION:
Why is the sea boiling hot, and do pigs have wings?

IMPORTANT NOTE for those struggling to present their first few articles, talks, or posters, and anxious to know the basics. The book is in two parts (Figure P.1 below). *PART I: BASICS* contains four chapters. One of the first three ('1. Writing and publishing an article', '2. Speaking about your work', and '3. Presenting a poster') should be sufficient to help you to get started: one learns best by doing. Chapter 4 ('Scientific authorship') discusses the rewards and some of the hazards of authorship.

PART II: IMPROVING contains chapters that may help you to improve: '5. Writing style', '6. Misused words and concepts', '7. Quantitative matters', '8. Treating errors', '9. Data interrelations', '10. Tables and figures: the evidence', and '11. Citing and referencing'. A Bibliography, List of references, and an Index are at the end.

I recall with gratitude the late Cyril Mummery who set me on this road and the late Tony Fogg who encouraged me along it.

I thank, and so may you, those who have helped me: Brian Moss and Dick Webster for constructive criticism (and uncounted improvements) throughout the book; Alexandra George, Peter Grubb, Steve Ketteridge, Andrew Leitch, Richard Nichols, Håkan Rydin, and Tony Walsby for equally helpful improvements to one or more chapters; Alan Crowden for encouragement in finding a publisher; and at Cambridge University Press, Vania Cunha, Dominic Lewis, Ilia Tassistro, and especially Anna Hodson, for easing the book into print.

Finally, I thank all the individuals in that shadowy muttering crowd of those whose ideas or practices I have absorbed without remembering or even knowing the source. May they forgive me.

The defects that remain are, of course, my own. I welcome corrections and other amendments.

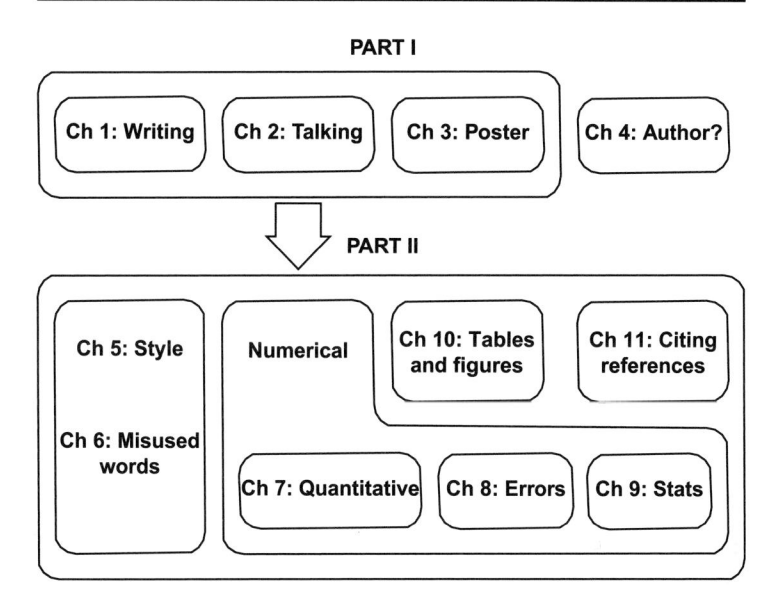

FIGURE P.1 Structure of this book. Chapters in Part II contain detail useful for the ones in Part I.

PART I
Basics

1

Writing a scientific article and getting it published

A piece of scientific work is not complete until the results have been written up and published. Planning and doing the work is often exciting; writing about it may seem less so but is just as necessary, and delivers its own satisfaction when you see the work in print.

Why bother to acquire skill in describing your work? First, as you *have* to produce reports then better to do it well than badly. Second, the number of scientific articles published in a year has been doubling roughly every decade or two, but the time that any one scientist can spend reading does not increase. If you write badly those who ought to know your work may not be willing to make the effort needed to understand it, and those who do try will suspect that someone who writes so badly may have been just as confused and incompetent in doing the work as they are in trying to report it.

Most scientific articles exist between the ephemeral and the eternal. If you write concisely and clearly, your readers may begin to look forward to reading about your work, your reputation will rise, and the useful lifetime of your article will lengthen. Good writing cannot convert bad science into good – you cannot make a silk purse out of a sow's ear – but you can ensure that your work is read.

If you are lucky you will have had an opportunity to talk about your work (Chapter 2) before needing to write about it: 'lucky' because talking concentrates the mind and forces you to express things you know that you know but have not (yet) put into words. If an opportunity has not presented itself then make one. A small audience or even an individual is all that you need.

Assume now that you have done original work whose results will, you judge, be of interest to, and perhaps excite, fellow scientists in your field worldwide. How should you go about constructing the article ('article' also known as a 'report' or 'paper', perhaps preceded by 'research')?

The standard scientific article

A few reports of scientific work may be accurately historical: 'We did this and then we did that, which we later discovered was a dead end so we abandoned it and tried something else instead . . .' This approach can be fascinating if a major discovery lies at the end. James Watson's (1968) controversial *The Double Helix* recollecting his contemporary view of the search for the structure of DNA is of this kind. In particular, Watson's account of Rosalind Franklin's part in the DNA work is widely thought to be unfair to her (though it was an accurate reconstruction of Watson's perception at the time). Rosalind Franklin died of ovarian cancer in 1958 at the age of thirty-seven. A corrective account of her life is in the biography of her (Maddox 2002).

But most scientific work, though it may be essential, is mundane and most scientists are unwilling – probably unable – to spend the time needed to follow in the work of others the twists, turns, unproductive directions, and illogical order that characterise much of the daily experience of scientists. Compare Watson's (1968) informal account with the formal publication of the proposed structure of DNA (Watson & Crick 1953).

The standard structure of a report of scientific work has evolved over 350 years or so with a single main purpose:

to convey information to its readers as clearly and simply as possible.

5 THE STANDARD SCIENTIFIC ARTICLE

Such a report will be historically accurate only rarely, when the work went in a straight (unbending) and strait (narrow) line from idea to publication. "Scientific papers in the form in which they are communicated to learned journals are notorious for misrepresenting the process of thought that led to whatever discoveries they describe" (Medawar 1963, 1969). Working scientists are familiar with, and mostly untroubled by, the fact that the report is a sanitised version of events.

There is no single 'correct' structure for a standard scientific report about original research, but there is a usual one – "a ritual liturgy" (Jacks 1961) – that scientists are familiar with and which they understand easily. After the Title, the main headings are: Summary or Abstract, Introduction, Methods, Results, and Discussion ('TSIMRD'). In the medical field most reports of randomised controlled trials follow the CONSORT (CONsolidated Standards Of Reporting Trials) format. Similar standardised structures for specific sorts of medical study are QUORUM, MOOSE, and STARD. Medical reports are generally more ritualistic than others (Peat *et al.* 2002) and are becoming even more so. An unusual example of the double-blind trial format is described (Smith & Pell 2003) when applied to the efficacy of parachutes.

The next few sections take you through the processes of writing such a standard scientific report. 'Listeners tacitly expect speakers to be informative, truthful, relevant, clear, unambiguous, brief, and orderly' (Pinker 1994) and readers will have the same hopes of your writing.

Begin by choosing

- your target journal ▼, and
- someone to represent your readers.

▼ A journal is one kind of periodical that librarians may formally call a 'serial'. In this book, unqualified 'journal' indicates a scientific publication recording accounts of primary research.

Look in your target journal for the specific requirements, arrangement of sections and formatting instructions, then follow these. Different journals have widely differing requirements (and the clarity of their instructions also differs widely). Following a specific format saves a lot of error-prone messing about later, and creates a good impression when the article reaches the editor of the journal.

Write for your chosen person. Someone in a related scientific field is usually a good choice, but if you are about to write a more general account then you would need to choose your target publication and representative of your readers differently, and to use a different structure for your writing. A newspaper article, for example, often begins with a specific instance – a human story – and uses that to develop a generalisation. That technique is unusual in a scientific article, though it may occasionally be effective in attracting attention. Whoever the person you have chosen to write for, make sure they are critical. If their first language is not English then so much the better, for science is international and many of your readers will have difficulties with English (though scientific English is a simplified version of the full language). You may then assume that your chosen reader is scientifically literate, but you will know that they are unfamiliar with rarely used words and with low-level technical details, so you will have to explain clearly and simply what you have done and why you did it.

Complete as many of the data analyses, tables, and figures as you can before beginning to write. If the work you are going to report was other than simple and straightforward you will probably find during writing that you need more analyses and resulting tables and figures.

Some busy bioscientists manage to write good first drafts on railway trains and at airports or on planes. But to begin with you will probably do better to isolate yourself from distractions. When you are ready to start writing, then hide where you cannot be interrupted. Silence telephones, lock the door if you can, and respond only to fire alarms. It is difficult to concentrate on writing for more than a few hours at a time, so expect to make several such 'retreats' before you have a complete first draft. But once you have begun the whole process keep it going. It is easy to be diverted into other activities and thus to lose momentum.

Table 1.1 lists the sections of a standard scientific report. These are generic headings that allow a reader to recognise what sort of material will follow. But it is usually good practice to make subheadings to identify specific topics within each section. Each of the four largest sections answers a question. This structure has evolved over three and a half centuries and is what scientists everywhere will be expecting. If you deviate from it without good reason you will make your readers' task more difficult than it need be.

It is possible that your work does not fit the standard patterns that journals recognise. If you suspect this then seek advice from an experienced colleague. Remember: you are writing for self-selected readers, and they will be able to understand you most easily if you follow a pattern they are expecting.

Now consider these sections in turn.

Title

Stephen Hales' 1727 masterpiece of plant physiology was titled: *VEGETABLE STATICKS, Or, An Account of some Statical Experiments on the SAP in VEGETABLES : Being an ESSAY towards a Natural History of Vegetation. Also a SPECIMEN of An Attempt to Analyse the Air, By a great Variety of CHYMIO-STATICAL EXPERIMENTS; Which were read at several Meetings before the ROYAL SOCIETY*. The title as originally printed uses eight different fonts, which I have not reproduced here.

Nowadays titles are much shorter. Just as a précis reduces a chapter to a paragraph, or a paragraph to a single sentence, so a title seeks to summarise the Summary in a single phrase or short sentence.

Make the title as short, striking and interesting as possible. More than 20 words is probably too long. You are competing for attention. You need to try to entice readers, as they scan the results

TABLE 1.1 The sections of a standard scientific article and the questions they answer

SECTION	READERS' QUESTION
PRELIMINARIES	
TITLE	What is the work about?
SUMMARY (or ABSTRACT) and keywords	What might interest me?
(SYMBOLS, ABBREVIATIONS)	
MAIN SUBSTANCE	
INTRODUCTION	**WHY** did you do the work?
METHODS[a] (MATERIALS / STUDY AREA)	**HOW** did you do the work?
RESULTS	**WHAT** did you find?
DISCUSSION	**SO WHAT** are the implications?
(CONCLUSIONS)	Remind me what you conclude
SUPPORTING MATERIALS	
ACKNOWLEDG(E)MENTS[b]	Who helped you?
REFERENCES (or LITERATURE CITED)	Where can I find work that you cite in support?
(APPENDICES[c])	Your evidence?
TABLES, with Captions (Legends)	
CAPTIONS (LEGENDS) FOR FIGURES (and PLATES[d])	
FIGURES (and PLATES[d])	
SUPPLEMENTARY MATERIAL (on a website)[e]	
Text, Tables, Figures, Videos	

[a] Some journals put METHODS after the DISCUSSION or even as part of the REFERENCES.
[b] There are two spellings, but a particular Journal will probably impose one. Some journals include ACKNOWLEDGEMENTS in or near the SUMMARY
[c] Some journals put APPENDICES before REFERENCES
[d] PLATES were most commonly photographic illustrations, but the 'Plate' designation is rarely used nowadays
[e] Many journals now allow you, if you wish, to submit extra materials that will be displayed on a website only. Here you may put, for example, a video that can show apparatus, or a dynamic event that cannot be clearly presented at all in the conventional way on paper.

of an electronic search or a journal contents, to nibble at your bait. In that light a title with an intriguing metaphor or allusion may work well. 'Climate change: the Devil's trap?' Usually however the title is more prosaic.

The title should, if possible, be understandable by any scientist. Most of the words in the title should be keywords – words that are used for searching. Start with a keyword, not with 'The' or 'An' or similar padding. Be as specific as you can, consistent with brevity. If the article is about a single species well known to your target readers then give the name: 'Regeneration of laurel forest in Greece' is likely to be more useful to a potential reader than 'Laurel forest in Europe'. If you have compared *Laminaria* and several other large brown seaweeds then move up a rung and put 'large brown seaweeds' in the title. Put the main topic first: 'Taxonomy of *Laminaria* in ...' will be noticed by different readers from 'Climate change effects on *Laminaria* ...'.

Give the strain number of microorganisms, if there is one, because different strains often behave differently. Leave out the authority for plant and animal names (you can put the authority in the main text) because such details clutter the title unnecessarily.

Avoid all except the commonest abbreviations: DNA is acceptable, AAFS (atomic absorption flame spectrophotometry) is not, except perhaps in an analytical chemistry journal. Spell out simple chemical names: 'methane', not 'CH_4'.

There are various types of title. The commonest is a descriptive phrase (without a verb) defining the field:

> 'Radar surveys of the Moon's surface and the "blue cheese" hypothesis.'

Another form is a full sentence posing a question:

> 'Why does the Moon behave like an ellipsoidal blue cheese?'

A third form is a bold declarative statement making a claim:

> 'Radar surveys show the Moon behaves as an ellipsoidal blue cheese.'

Declarative titles do identify what the author thinks will interest readers, but although some journals tolerate or encourage such titles ('Radar surveys *show* ...'), others consider them bad because they seem to be trying to force a verdict on the jury before the trial has begun.

As you go further with the writing you may get new inspiration and can improve the title.

You can suggest, or may be told, to supply a 'running head' or 'short title' of fewer than 50 or so characters ('space' counts as a character) to appear at the top of alternate pages in the printed article. This may be a convenient place to put the English name ('brown seaweed') of an organism you named in the title in Latin.

Summary or abstract or (rarely) synopsis, and keywords

Make the Summary as short and clear as possible. Many journals restrict you to no more than 2% to 5% of the total length of the text or to 250 words or so. Every word must pull its weight. It may help to have up to half a dozen numbered sections in the Summary. Some journals specify the headings you must use in the Summary; others leave it to you to decide the format. Many of these prescriptions are fairly recent and reflect editors' recognition of the importance of the Summary; their variety shows the difficulty of deciding what a Summary should contain. For example, at least one journal asks for 'highlights' as well as the Summary.

What is the difference between a Summary and an Abstract? The *Shorter Oxford English Dictionary* gives no help: it defines 'abstract' as 'an abridgement or summary'. If there is a difference it may be that one is a précis, giving equal weight to all parts of the article, while the other is selective, concentrating on

- the purpose of the work (but only if not obvious),
- how you did it (but only if not obvious),
- what you found, and
- what you infer or conclude.

Opinions differ, however, on which of Summary and Abstract is précis and which is selective, and on whether you need the précis or the selective form. If the journal gives no instruction then prefer the selective version that concentrates on results and inferences, because it makes better use of the space (and of readers' time). What the editor of the journal calls this section scarcely matters. The fact is that it is here that you must try to entice some of those attracted by the title to scan the rest of the article.

The Summary or Abstract is what will probably get into electronic databases. For example:

> 'Side-scan radar images of the Moon's surface, in the beta band, show that: (1) the surface deviates no more than 30 m from a perfect oblate ellipsoid; (2) beneath a 1-m-thick rock shell the Moon is of the same consistency as Roquefort cheese.'

Notice that you can understand most of this, and strongly suspect that it is nonsense. That is what makes it a good summary. The Summary may contain a few technical terms ('side-scan images', 'beta band') that will not be familiar to all scientists. It should contain specific numerical values where relevant: some readers may cite your work in support but never bother to read the full article – reprehensible but common. If the Summary is all that they are going to read it had better contain the chief points from the full article.

But (because it will be copied on its own into databases) the Summary should contain no citations to references or table or figure numbers, and should avoid all but the commonest abbreviations. It should concentrate on short unambiguous statements of fact, and contain no vacuous phrases (such as 'The results are discussed' or 'It was shown that ...' or 'Experiments were made to ...').

It is surprising how many articles wrongly include material in the Summary that does not appear in the rest of the article.

Most journals will ask you to suggest 'keywords' (which may be short phrases) that are printed as part of, or just after, the Summary or Abstract. These words are put into databases and are used in searches, usually to supplement the words in the title: you should not need to repeat 'radar', 'survey', 'Moon', or 'green cheese' from the title as keywords. Some journals do, however, ask you to include important words from the title because a few search engines do not look at titles. And some journals no longer require keywords at all relying instead on the major web search engines that scan the titles. Some authors try to bring their work to wider notice by including keywords that are nothing to do directly with their article: 'cheesemakers' for example in the Moon survey article. This seems to me to be a dubious and probably ineffective practice.

The Title and Summary owe something to journalism in that you are trying to attract attention. But do not get carried away. Avoid inflated claims and exaggeration.

Symbols and abbreviations

Most bioscience articles move straight from the Summary to the Introduction. But before doing so, ask yourself two questions. Are you going to use numerous symbols to represent variables and parameters? Are you going to use several unusual abbreviations (such as 'bv' or 'L.D-Bc')? If so, and particularly if these are to be used sporadically throughout the text, then it will help readers if you list them all here with their definitions.

If you do need such a list, put it here at the start of the article, where readers will remember it exists, not at the end where they will probably not see it in time for it to be useful.

Introduction

Here is the place to explain WHY the work you are about to describe was started, and what its purpose is. You need to review earlier work selectively and briefly – five to ten references should be sufficient to set your intentions in context and to recognise the important work of those who have gone before you. Then state the purpose of your own work. The Introduction generally narrows the focus from 'What is already known' through 'What we need to know' and perhaps 'Why it is important' to 'The purpose of this article is ...' In many cases two or three paragraphs on a single page are sufficient. A modern fashion is to cast purpose in the form of one or more hypotheses. This can be a powerful way of organising work, but is not universally suitable. There are three common elements in most scientific articles:

- survey or observation, with or without a prior overt hypothesis;
- experiment, often but not always with a hypothesis;
- analysis, often mathematical, with or without a hypothesis, using logic.

Some scientists have a preference (or, at least, are best known) for one of these three elements: for example Galen and Darwin for observation (with inferences), Boyle, Hooke, and Mendel for designed experiments, and Newton and Einstein for mathematical analysis. Many scientists use more than one – sometimes all three – of these approaches (Newton made experiments too) and the proportions of these three elements in a single article differ widely ▼.

> ▼ The core of the scientific process is to have an idea then to test its truth: 'Is it possible? Is it actually so?' Often the idea is framed as a hypothesis, and the test is commonly a designed experiment, usually with replication. Sometimes, however, the work is to answer the child's question 'What happens if ...?', and this approach is perhaps the most likely to lead to unexpected discoveries. Some areas of science, such as astronomy and cosmology, cannot make the designed experiments they would like to, so they observe the results that Nature herself provides, though She seems to have missed the celestial course on 'Experimental Design'. These observations may be made to test a specific hypothesis ('are galaxies distributed randomly in space?') or the hypothesis may follow an observation (pulsars; the correlation of infant deformities with the thalidomide taken by their mothers, and that between smoking and lung cancer). Early work to describe, name, and group stars, and organisms, was (and still is) essential, and may be the basis for studies of

> evolution of stars or organisms. Descriptive work can be just as scientifically important as experiments. The discovery of unusually large concentrations of iridium at the geological K–T (Cretaceous–Tertiary, now K–Pg for Cretaceous–Paleogene) boundary, coinciding with the disappearance of large dinosaurs, led to the suggestion that a large meteor had hit the Earth and caused, or at least contributed to, the extinction of large dinosaurs. And work on the Burgess shale, formed in early Cambrian times after the first multicellular organisms had appeared, revealed more than 20 basically different arthropod body plans, none obviously superior to any other, of which only four survive (including insects, spiders, crustaceans, and trilobites). This led to the suggestion (Gould 1989) that chance, acting in a field defined by physics and chemistry, has had a major rôle in evolution. This view was later disputed (Conway Morris 1998) with an alternative view that in fact there is only a limited number of possible developments. Both views arise from careful palaeontological descriptive work.

If you had more than one purpose then arrange them in a logical order, not necessarily the order in which you did the work. A concise Introduction might look like this:

> 'The first astronauts to land on the Moon noted (Oldeman 1974) that the surface was surprisingly smooth, and concluded that the appearance of mountains in photographs taken from Earth must be an artefact or illusion. Radar survey techniques have since improved by three orders of magnitude, and now allow a precision better than 10 cm. We describe here their use in the first complete radar survey of the Moon's surface.'

Note: claim 'the first' only if you are sure!

Very often you need to say that there is some sort of conflict, or an unexplained observation that you thought you might be able to elucidate, or a theoretical prediction that you wanted to test, or just that you wanted to explore a hitherto uninvestigated or under-investigated natural phenomenon. The last sentence in the Introduction commonly implies: 'Here is what I did to answer this question'.

At the end of the Introduction some authors state the main results that are to come. If the Summary is well written there should be no need at all for such out-of-place repetition in the Introduction.

Space in journals is expensive (hundreds of €/£/$ a page) and scientists' time is limited. Too often an Introduction is submitted as it was in a thesis: as a fully referenced review of a whole field. There are special review journals for that sort of thing. Editors of primary research journals sometimes invite a review. Discuss with the editor before writing if you wish to submit one. Editors of specialist review journals usually invite reviews from those with an established reputation in the field of the review, but you can make an offer to them too.

Concentrate on citing mainly the earliest, the most comprehensive, the most recent, or the work that is 'best' in some way. It is lazy to cite only the first reference you can lay hands on. Would you like to see your early important work substituted in 10 years' time by that of some Johnny-come-lately? Do not feel obliged to mention weak articles or ones that add nothing useful or are only marginal. Readers who have got to the Introduction are hooked, and they join the editor in urging you to be concise. Most journals are restricted to a set number of pages each year. Popular journals (sometimes called 'good journals') get perhaps five to twenty times as many articles submitted as they can publish. Many journals restrict you to five or six printed pages for an article, or charge for extra pages. If your Introduction is clear and concise you increase the chance that your article will be accepted, reduce supplementary charges, reduce the chances of error, will have less work in revising the article, and cause more people to read and understand what you have written.

By the end of the Introduction your readers should, metaphorically, be moving along at your elbow saying 'Yes. I can see why this was worth doing. Now, HOW did you set about it?'

Materials, study area, and methods

Articles that deal with observation or theoretical matters may not need a Methods section, but nearly every other sort of article does. The heading 'Methods' often covers more than you might expect: background (for example sites, organisms), methods (in the strict sense) used to make measurements, and details of experiments ▼. These components may be grouped under subheads, or may be given their own major heading. For example, ecologists often need to set their work in context: they describe the salient features and climate of the site(s) they worked on, so one gets a section headed something like 'Study site'. Many projects are made to discover something about a particular organism, so there is a first section headed 'Materials' where the organism and strains are named followed by a subsection headed something like: 'Culture of *E. coli*' or one in which plasmid construction and propagation are described.

> ▼ Some subjects (chemistry is an example) may put the whole of the Methods near or at the end of the article, and journals such as *Science* and *Nature* put them in endnotes (often too sketchy to be useful). Most laboratory-based bioscience journals put Methods after the Introduction, but microbiologists usually put details of the experiments in (sub)sections of Results. Here I deal with procedures and experiments together as Methods. Look in your target journal to see what its custom is.

If it is not obvious, then begin each subsection of Methods with a brief summary to explain how the materials, procedures, and experiments you are about to detail relate to the objective of the work just described in the Introduction. 'Profiles of gas concentration, to allow calculation of fluxes, were measured by ...' Your readers will not then be mystified when you describe how you operated a portable mass spectrometer.

The ideal is to give enough specific detail that someone expert in the field could repeat the essentials of your work. Explain, if it is not obvious, why you chose a particular method, what its advantages are, and any difficulties there may be in understanding results got with it.

Group the methods proper by what they do, not by the pieces of work they were used in. Many readers will skip most of the Methods section on first reading, particularly if the methods are standard ones. Only later may they need to find out exactly how you made some particular measurement. They can locate the description most easily if it is in a group with a common purpose. You may, for example, have a section headed 'Chemical analyses' and that is where I expect to find how you measured pH.

For standard methods you may just cite a reference, but do also give the principle, in outline at least: not just 'Humification was measured by the method of Bloggs (1998)' but 'To measure humification we first extracted samples in boiling 8% NaOH, then measured absorbance at 450 nm (Bloggs 1998)'. For the complex standard protocols common in molecular biology you *must* detail any deviation from the cited source.

Try to avoid trade names: 'Perspex', 'Plexiglass', 'Lucite', 'Acrylplast', and 'Acrylite' are all the same chemical polymer of methyl methacrylate, and are 'acrylic glass' almost everywhere. Some authors name specific commercial instruments. This may help readers if the equipment is large or unusual, but there is little point in advertising the maker of the commonplace pH meter or spectrophotometer that you used unless you think the measurements made with it may be questioned by some readers.

If you devised a new method (or used an unusual one) you will probably need to explain the context and principles of the method, and give more details than are necessary for a standard method – sufficient to allow a reader to try the method.

Write the procedure descriptions so that your readers understand what you did. You filtered the suspension, good, but did you then use the filtrate, or what had been retained on the filter paper? Imagine you are watching a film of yourself at work, and describe processes in their natural order. Be suspicious of 'after' in a procedure description. Not 'the suspension was filtered after boiling for 20 minutes' but 'the suspension was boiled for 20 minutes, then filtered, and the filtrate concentrated ...'. Include specific times, temperatures, concentrations, volumes, and similar details.

If your work involved both non-trivial measurements and experiments then make one section for the measurement procedures and another (later one) to describe the experiments and their details. (Use the subheading 'Experiments' not 'Experimental': headings should contain at least one noun). If suitable, use a different subheading for each experiment, and use the same subheading name for a specific experiment in both Methods and Results. If there were several experiments it may be helpful to summarise their salient features in a table.

In Methods you should also name any computer programs and databases you used, and describe any unusual statistical or other numerical analyses – if it is not in a 'Statistics for XXX Students' text then it is probably 'unusual'. Give a brief description of what the unusual method is to do and perhaps of how it does it. Just what is 'Hibbert sampling' or 'neural analysis'? The editor may strike some of this out, but is more likely to be glad to have a brief explanation.

Finally, a common mistake is to include results in the Methods section. Save for the first part of Results the demonstration that your ingenious new method does what you intended.

By the end of Methods your accompanying reader should be saying 'Fine, fine. Seems good. I can see how you set about things and it all looks promising. Now, WHAT did you discover when you actually did the work?'

Results

Here is the heart of your work where you tell your readers what you found. The main purposes of the Results section are to present the results, to draw attention to the important ones, and to explain why some features that may look important are really not so. It is usually helpful to use subheadings to separate blocks of results.

Avoid the temptation to describe again in words what is already clear in a figure or table. Use the text to *comment* on the results, particularly when the reliability or significance of a result is unclear or even misleading.

'The results of the second survey in the beta band show that no point deviates from an oblate ellipsoid by more than 30 m (Figure 7), but this result conflicts with all previous surveys using other methods.'

Results should usually be presented either in figures or in tables, but rarely in both (if essential, put the numerical detail in an Appendix, or into an electronic database). Most scientists find it easier to get the message from a figure than from a table, except perhaps for multivariate results or where the information is text. It is worth spending time choosing the best form of presentation: it can make a big difference to clarity. Chapter 10 ('Tables and Figures: the Evidence') gives details.

Most journals require you to number the tables in one series and the figures in another: Table 1, Table 6 and so on, and Figure 1, Figure 6 and so on. Each table *must* have an integral explanatory legend or caption, while the equally mandatory explanatory legends or captions for figures should, unless the journal specifies otherwise, be collected on a separate sheet headed 'Figure captions'. (Why? Long ago, figures went to the blockmaker, while captions went to the typesetter.) In ideal circumstances each table and figure (when attended in public by its caption or legend) would be intelligible to someone knowledgeable in the field if they saw it in isolation. Imagine making a slide of the table or figure with its caption, and showing it in a talk. Would the members of your audience be able to understand it before you spoke? (I ignore here the need to simplify most tables before showing them in a talk.)

Even if each table and figure is not understandable in isolation the Results section as a whole should be. You must include signposts so that your readers understand the circumstances in which each group of results arose.

'The results in Figure 5 of the second survey using the beta band show that ... '

If your Methods section and the figure caption are clearly written this is all the linking that is necessary, but this short bit is vital.

You may also need to include a brief discussion to set the scene for the results of the next bit of the work.

'This surface uniformity can be explained only if the interior is surprisingly fluid, with a viscosity not much greater than that of cold treacle. We therefore used the beta band scans to calculate the viscosity exactly using the aperiodic technique already described. The results are shown in . . .'

Include only as much discussion as is essential for clarity: it is easy to wander off into detailed but premature inferences and speculations thus mixing Results and Discussion in a confusing way.

If your results are voluminous you will not be able to include them all in the main article. Many journals now maintain an electronic repository for supplementary results and other types of material, such as videos ('Supporting information'), and there is an increasing number of similar freely accessible repositories for raw datasets that will never be published in any journal. Be careful to avoid misleading readers when you describe what you have chosen to publish: the best is not 'typical'.

It is usual to omit 'failed' experiments – ones that went wrong in some technical way during execution – but hesitate before excluding a result simply because it was not what you expected or because it is some way away from a trend shown by all the other results. If it is wildly erratic – isolated and, say, five standard deviations away from the mean – you may choose to suppress it, but you may thereby be throwing away the most interesting result of all. Polyethylene was discovered in this way in 1933, as the result of a 'failed' experiment (later found to have resulted from a leak allowing tiny catalytic amounts of oxygen to enter the reaction vessel). Commercial production began on 3 September 1939, just in time for polythene's high electrical resistance, low density, and consequent light weight, and mouldability to make airborne radar practicable. The discovery of the ozone 'hole' over Antarctica (Farman, Gardiner, & Shanklin 1985) is a similar example.

Often a few values may seem to be aberrant but not so wildly that you can suppress them with a clear conscience. One solution is to report these values, but then say that you are going to omit them from statistical analyses and to ignore them in the Discussion. Beware of 'rules' that permit the removal of a value that is at least one standard deviation away from the next nearest value. Repeated application of such a rule may be effective in misleadingly removing a long tail completely.

In quantitative work you should give some idea of the exactness of replication. 'The mean value for the untreated plants was 92 mm and for the treated ones was 99 mm.' is not, in itself, of much use. If the standard error of each mean were 1.5 mm these differences would be interesting, but if it were 27 mm then the means have low precision and there is no clear indication of any difference (Chapter 8 explains why). Usually you will need to use at least some simple statistical treatment. All statisticians would tell you that you should have planned the statistical treatments, with advice if necessary, before you did the practical work. If you did not then there is a big risk that you can draw no useful conclusion (a postmortem may tell little except what the experiment died from). Almost as wasteful is to find that you have made far more measurements than were necessary.

Even if there is a statistically significant difference you should consider whether it is scientifically important in the context of your work.

By the end of Results the reader at your elbow should be saying 'Gee, wow. How interesting. I can see what you discovered. And I can understand now why you think that small difference in Figure 5 is critically important, and why the much bigger one in Figure 7 does not actually mean much. SO WHAT are the implications of all this?'

Up to this point you should have brought your readers along with you, nodding (notionally) in agreement most if not all the way.

Discussion

This is where you may, if you wish, spread your wings and be speculative, provocative, or even controversial. For this reason, if no other, never have a mixed 'Results and Discussion' section.

In the Discussion readers will be hoping to find your interpretation of what has gone before. Almost inevitably some of what you write here will be speculative, but you also need to be persuasive. Readers who have accepted your methods and results without much resistance are likely to be more alertly critical of your Discussion. Remember that you will have to defend your views. Try to anticipate criticisms. Relate the Discussion to the purpose(s) of the work listed in the Introduction. Use subheadings to separate topics. Weakness in the Discussion is one of the commonest topics of criticism by referees. If your work has caused you to change a view you have previously published then do not be afraid to say so. Science is a series of successive approximations to the truth and it advances by correcting wrong or incomplete views. The economist Maynard Keynes when accused of changing his mind is said to have replied "If the facts change, I change my mind. Pray sir, what do you do?"

What is your main finding? How reliable is it? To what extent do your original hypotheses now seem implausible? Are they still plausible only because the scale of your work was too small, given the variation, to allow you to reject them? Do you now have a new hypothesis, and how far are the results consistent with it?

If the work has not achieved its original purpose then explain why not, but do not run on and on about errors. A clear-cut negative result may be valuable but a heavily qualified mess is not: "Truth comes out of error more readily than out of confusion" (Francis Bacon, *Novum Organum*, 1620). Compare your results with the results of others. Are yours in some demonstrable way 'better' than what was known previously? Often the Discussion broadens the focus as it goes on, from the particular to the more general (thus reversing the initial narrowing in the Introduction).

If the problem in the Introduction was well posed and the Results were clear-cut there may be little to be said in Discussion. Don't be afraid of having only a short Discussion: it may have greater effect than a long one because its few important points are unobscured by waffle.

Conclusions

Decades ago this is where the Summary was put, before it was moved to the start of the article for the convenience of readers. Some journals require a Conclusions section, and some authors write one without prompting. If your article is unusually long, or complex, or has required intellectually difficult argument, then there is a case for recapitulation of your main findings here.

Often, however, the so-called conclusions are no more than a last-ditch attempt to explain again (or more clearly) what has been left in confusion by the poorly written earlier material. If you find yourself wanting to write a Conclusions section then consider whether or not it is a sign that you are dissatisfied with what you have already written. If not, the impulse may be no more than the recognition that it would be good to end the Discussion on an upbeat note by turning round at the top of the hill and gazing back at the valley you are about to leave. Authors of research articles in the *British Medical Journal* are encouraged to specify, for printing in a framed box, what was known before they began their work and what they think they have added. This is a salutary (health-giving) discipline and, as part of the ordinary text, may be a good way to finish the Discussion.

If you do need to have a Conclusions section then use it to try to pick out a 'take-home message'.

> 'As a result of these new radar techniques and some unusual mathematical analyses we are driven unavoidably to the conclusion that the Moon, remarkably and as far as we know uniquely, behaves dynamically as a giant blue cheese would.'

Acknowledg(e)ments

'Acknowledge' means to recognise that something is so. What you really need to do here is to *thank* people and institutions for help of various kinds.

There are fashions in Acknowledgements, ranging along: active ('We thank ...'), neutral ('We are grateful to ...'), passive (' ... is thanked ...), mere observer ('... deserve thanks for ...'), factual

('... helped with ...'), to supplicant ('... wish to thank ...'); and none. Avoid 'wish to thank': here at last is your opportunity. Take it. Thanks are often reserved for individuals, while corporate bodies may be acknowledged factually. The individuals who have helped you are likely to be most appreciative of the active ('We thank ...'). It may be wise to get permission to acknowledge them from those who have given you professional advice (lest they disagree with what you have done with the advice and wish to avoid being linked to it).

Keep your thanks as short as possible consistent with recognising your obligations. Not 'I am eternally grateful to both my parents and wish to thank them jointly and severally for hospitality during the fieldwork and for their generous support over 22 years ...' but 'I thank my parents for housing me during the fieldwork; E. Blofeld and J. Smith ▼ for advice about cats; and Queen Mary University of London for financial support'.

> ▼ 'J. Smith' conceals Lady the Reverend Dame Professor Janice Smith CBE, FBA. Opinions differ on whether or not one should include given names, academic, professional, and other titles and distinctions in Acknowledgments. Do what you think the person you are thanking would like.

Use semicolons to string your thanks together. Some grant-giving bodies and some journals ask you to specify any grant number in the Acknowledgments.

If the article has more than one author then, if you wish (see Chapter 4), this is the place to record briefly which of you is responsible for what parts of the work: 'AB conceived the idea and wrote the first draft of this article, AB and CD designed the experiments, and CD carried them out and analysed the results.'

References (or literature cited)

You need to cite other documents (for example journal articles, books, reports) in support within your text, and to give the full bibliographic reference for each citation at the end. Everything cited must have a reference, and there must be no reference that is not cited.

There are several systems and formats for citing a reference, several ways of ordering references, and uncounted formats for the references. Your target journal will have made its choices but may leave you to infer what these are from a recent example of the journal. Details of the common systems are in Chapter 11. It is *your* (admittedly tedious) duty to put the citations and references in the correct format, and to list them accurately.

Some journals forbid citations to 'unpublished results' and 'personal communication' because the evidence is not available to readers, and such citations have been abused. Even if the journal permits them they may specify the form of words to use.

Tables and figures

You need to show enough of your results to support the statements and claims you make, but in many pieces of work that will be only a fraction of what you have done. Choose representative examples: the best is not 'typical'. If you show the best then be honest and say so.

Number the tables consecutively in the order that they appear in the article: Table 1, Table 2, and so on. Number figures similarly as Figure 1, Figure 2, and so on. A majority of journals use an initial capital 'T' for Table and 'F' for Figure, but some abbreviate 'Table' to 'Tab.' and 'Figure' to 'Fig.'. See what your target journal does.

As with citations of references, ensure that all tables and figures are (correctly) referred to in the text and, of course, that all tables and figures referred to in the text are actually present. It is easy to decide to omit or to add a table or figure and then overlook some of the necessary changes in numbering in the text.

The commonest types of figure are graphs, line drawings (examples are apparatus, maps, flow charts, organisms), and photographs (examples are experimental sites, chromatograms, gels). Photographs used to be expensive to reproduce but with modern printing methods are no longer so. Colour is still expensive when printed. Some journals do not charge for colour but others may expect you to pay a substantial amount towards the cost of a single image.

Detailed advice about constructing tables and figures is given in Chapter 10.

QUESTION:
Can you think of an actual or hypothetical account of work that does not fit the standard pattern described above?

Supplementary materials

Supplementary materials are optional. If you do submit them then they will be displayed on the journal website only, and not on the paper version of your article (if there is one). They may include text, tables, figures, references, and short videos. These last can be a powerful aid to understanding apparatus, methods, and dynamic events, particularly short-lived ones. They are an undoubted, and increasingly important, advantage of electronic forms of publishing.

The opportunity to present details in the 'Supplementary materials' section has another advantage. The main article can then be kept free of fine detail. This makes the main article shorter and thus it communicates a clear message more effectively.

Latin names of organisms

Latin names of genus and species are italicised thus: *Homo sapiens, Mentha aquatica* (water mint), *Escherichia coli*. The genus always has an initial capital but nowadays the specific name never does even when it derives from a person's name: *Chlorella pearsalli* (a unicellular green alga), *Dendrochilum flos-susannae* (a very small orchid), *Haloquadratum walsbyi* (a remarkable square archaebacterium living only in saline waters). The generic name carries its initial capital even in the middle of a sentence. Latin names of taxa lower than species (for example subspecies, form, variety, cultivar) are also italicised, but the introductory 'ssp.', 'f.', 'var.', 'cv.' is not: *Panthera tigris* ssp. *altaica* (the Siberian tiger). In a handwritten document underlining is used in place of italics thus: <u>Homo sapiens</u>. The rules for names below the rank of species are more complicated than given here, and are group-specific. In particular, those for microorganism strains are beyond the scope of this book.

Names of higher-order taxa (Family, Order, Class, Phylum, Kingdom) are not italicised, so: Poaceae (grasses), Insecta (insects). If a generic name is used in an anglicised form then it too is not italicised, but usually retains its initial capital: 'At sunset the Laminarias were an almost luminous brown'. Anglicised family names are an exception: members of the Family Hominidae become hominids.

In general, at the first mention of an organism in an article its generic name is spelled out in full and the name of the authority who gave that name may be added in abbreviated form. The old practice of including the authority in the title of an article when an organism is named there has almost died out, and some journals no longer require the authority to be named at all unless it is in dispute. After the first naming in full, where there is no possible confusion, the genus name is reduced to its initial capital: '*Chlorella pyrenoidosa* Chick.' at first and *C. pyrenoidosa* thereafter unless the generic initial would be the first character in a sentence. But if you were speaking about *C. pyrenoidosa* and *Ulothrix zonata* you would probably call them, after their first full specification, by their generic names '*Chlorella*' and '*Ulothrix*' rather than '*C. pyrenoidosa*' and '*U. zonata*'. Even stronger would be the compulsion to refer to the two thistles *Carduus nutans* and *Cirsium vulgare*, whose generic names have the same initial letter, as '*Carduus*' and '*Cirsium*'. That may be the better way to write as well.

Recall how strange it sounds when a journalist writes or speaks of 'C difficile', not knowing that the genus is *Clostridium*, or simply 'leylandi' not knowing that the aggressive conifer is the hybrid *Cupressus* × *leylandii*. Equally odd are journalistic Victorian-sounding references to 'the chlorella ...'.

What journal should I choose, and how many articles should I write?

Job-seeking, promotion, research assessments, and even survival in a job often depend on your publication record. The wisest institutions are selective, and the more so the further you advance: applicants for a professorship may be asked for their ten 'best' publications only. Unwise institutions just count the number of publications. Between these extremes are those that consult a Citation Count (CC) and may look at the Impact Factor (IF) of the journals in which you have published. What are these bibliometric measures?

Both start from a database called a 'citation index' (CI) containing details about published articles. For each is recorded at least the title, the author(s), the journal in which it was published, the other usual reference details (see Chapter 11), the Summary or Abstract and (crucially) the references cited in the article. The best known CI is *Journal Citation Reports* (JCR) maintained by Thomson Scientific (TS, a section of Thomson Reuters, and developed from Eugene Garfield's 1958 *Current Abstracts for Chemists*). This CI scans about 10 000 journals – the number increases with time – which are a subset of all journals, chosen by TS as the most important current vehicles of formal printed or, increasingly, electronic communication. The Index allows you to do such things as tracking an idea or an author back through the years, following an article forwards, or simply looking for all there is about a keyword. Author Citation Counts and journal Impact Factors are by-products. Competing databases are developing. Scopus scans roughly the same number of refereed journals as JCR. Google Scholar embraces the whole web, without selection or exclusion, and thus has wider scope than IF.

For a Citation Count (CC) you select a particular article, and are shown all the (necessarily) later citations of it, excluding self-citations (citations in later articles by the authors of the article being assessed). Because authors are often sloppy with their references you may have to look for inaccurate versions of author's name, journal title, year, or page numbers as well. Roughly 20%, though falling, of published scientific articles are never cited at all (Larivière et al. 2008, Wallace, Larivière, & Gingras 2009), and many more are cited (other than by the authors) only once. There is a general belief that a much-cited article is an influential one (or is so badly wrong that many other authors have published a rebuttal). But it would be foolish to conclude that all little-cited articles (and their authors and contents) are of low quality or unimportant. Some articles are 'sleepers' whose importance is not realised for many years. The citation half-time – the number of years from the year of publication to include half the total citations – depends on the subject. In biochemistry, molecular biology, microbiology, and genetics it ranges from 4 to 6 years; but in physiology, aquatic biology, ecology, morphology, and systematics it is often more than 10 years, and in mathematics more than 30 years.

The 'h' index (Hirsch 2005) is an alternative to a CC. It is the number, n, of articles with a CC $\geq n$. An h of 12 means that 12 of your articles have been cited at least 12 times. This index emphasises influence rather than quantity of publications, but is of little use for those in the early stages of a career. It disadvantages those who publish a small number of very highly cited articles. It also strongly favours those who work in groups and those in popular fields.

Impact Factors (IF) seek to measure the performance of the journal, rather than the individual author. For each article in the target journal JX in a given year IF counts all citations in all journals in the Thomson database (including self-citations) during the last two complete years, then divides by the number of articles published during the same two years in JX. This gives an average number of citations per article in JX.

One can guarantee that any system of performance measurement will be at least partially subverted by the ingenuity of those it is applied to. IFs have several and severe defects in construction and use. Let us count the ways.

1. There is a poor and, since 1990, decreasing correlation – from a high of +0.3 – between CC and IF because the subsequent CC of articles in a journal is strongly skewed to low CC (Seglen 1997, Lozano, Larivière, & Gingras 2012). Only a few articles contribute substantially to the journal IF, so the great majority of articles in a journal with high IF are referred to infrequently. This alone should have precluded the use of IF in the assessment of individuals.
2. Unscrupulous editors have boosted their journal's IF by adding to articles they are about to publish citations to articles in their own journal, and by writing numerous editorials that refer frequently or only to recent articles in their own journal. These citations count toward total citations while editorials do not count toward total articles published.
3. A different abuse arises when groups of journals agree to artificially increase citations to articles in other journals in the group ('stacking': Noorden 2013).
4. Even worse is the editor who invited authors of submitted but not yet accepted articles to add citations to recent articles in the journal: a suggestion understood by the authors as affecting the chance that the article would be accepted.
5. A single event may sometimes disturb an IF spectacularly. A free program is widely used in crystallography. A review article published in *Acta Crystallographica – Section A* included the suggestion that the article might be used as a general literature citation. Authors of some 6600 other articles acted on this suggestion the following year and the journal IF rose 20-fold from 2.5 to 50.
6. The IF also depends on the database used, and non-trivial revisions in the IF of some journals have followed corrections in databases.
7. An IF judgement based on a CC half-time of 4 years may be long enough in a motorway outer lane such as molecular biology, but is too short for most of ecology moving more slowly in the inner lane. And a densely populated field will produce more publications, though perhaps not proportionately more influential ones, than will a sparsely populated one. The IF of the top-ranked

31 WHAT JOURNAL SHOULD I CHOOSE?

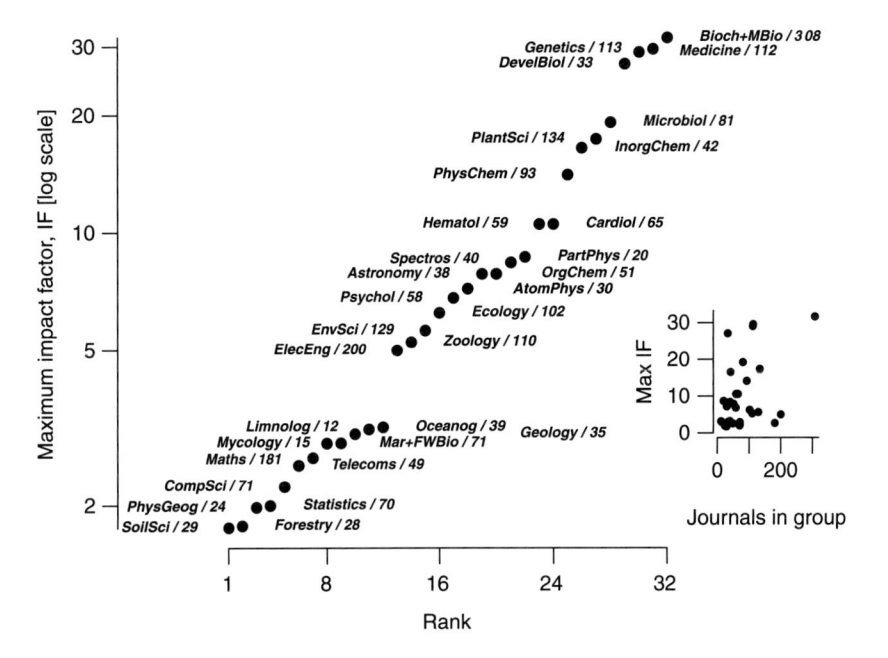

FIGURE 1.1 *Main graph*: log scaled maximum journal impact factor (IF) in 2001 in each of 32 subject groups. Subject group names alternate to left and right for readability. The integer after the group name is the number of journals in the group. *Inset*: IF in relation to number of journals in the subject group, showing no clear relationship.

journal in different subject groups is distributed linearly on a log scale from over 30 to about 1, as Figure 1.1 shows. (The inset to Figure 1.1 dispels the suspicion that the maximum IF may be a consequence of more journals in a subject group, but the number of journals is only a rough measure of the size of the community they serve.)

For these reasons one should never use IF to compare journals or workers in different fields, or, even worse, to assess individuals. These and other points were made in December 2012 in the Declaration on Research Assessment (DORA) that has since been widely supported by numerous organisations and individuals. But will all the members of a Selection Committee know this? Will they realise that an article in a journal with high IF for its group may have little merit, but that one in a journal with low (or no) impact factor is not necessarily poor or of little influence ▼? And will they realise that a journal impact factor is a poor surrogate for the citation count of a particular article, and that neither is any substitute for reading the article itself?

> ▼ I published a short article in an obscure journal that is not even in the TS Citation Index (Clymo R.S. 1967. Movement of the main shingle bank at Blakeney Point, Norfolk, *Transactions of the Norfolk and Norwich Naturalists' Society* 21: 3–6). Fourteen years later, to my astonishment, it was reproduced in one of the *Classic Papers in Geology* series. But remember that 'One swallow does not make a summer.'

A newer attempt to assess journal influence is the '$h5$' index operated by Google. A journal's $h5$ is the largest number of articles, n, for the entire journal that have been cited at least n times over the last five years. It is better than IF in that Google scans the internet as a whole for its database (Google Scholar), not just the Thomson database, and spans five years not just two. But it too should not be used to assess individual scientists.

The view is widespread that Impact Factors and, to a smaller extent, Citation Counts may prove – are proving – poisonous to the scientific enterprise (Colquhoun 2007, Lawrence 2007). The misuse of IF by some administrators and university 'managers' to allot funding, or to refuse promotion, or to justify redundancy, is seriously disturbing.

Several other bibliometric measures of individual and journal influence exist (Webb 2008). All such measures have several and severe defects and restrictions.

How then does one choose a target journal? There are several stages and various criteria. First, Table 1.2 shows the sorts of scientific article and the sorts of publication in which they may appear. Here we are concerned with an article that is to record original work – part of the primary record. Most authors aim for a refereed journal (in which the editor has sought advice from

TABLE 1.2 Sorts of scientific article and of publication. Prestige decreases downwards, but work of high quality may be found in any of the categories in this table

Primary account Large or wide circulation journal, refereed Journal with smaller circulation, refereed International conference,[a] refereed	Review Dedicated review journal Chapter in an edited book
'Grey literature' Thesis[b] Conference, not refereed Departmental or other Report	

[a] See warning in the text.
[b] The American and British thesis is a self-contained work, possibly supplemented by published articles. Some journals will not accept citations of such theses, considering them to be 'unpublished'. In many other countries the thesis is a bound collection of already published, or accepted, or (to stretch a point) submitted articles in journal format. The whole is stitched together by a written commentary.

experts in the field) and preferably one with a wide circulation, because these journals will be more widely available and read, carry greater prestige, and may give the author more help than other journals. Readers expect the refereeing process to ensure at least a threshold quality of work and presentation.

Second, you have to decide how interesting your work is likely to be to a wider public. This is more difficult. A few journals accept work about any aspect of science. Most are narrower. The narrowing may be, for example, by subject area (biochemistry, genetics, ecology), to differing degrees by organism (microbiology, botany, zoology, diatoms, mammals), by method (experimental botany, biological statistics), or by habitat (Antarctic, forest, peatland). A journal with a country name in it (*Australian Journal of Botany*, *Annales Zoologici Fennici*) may have wide scope. One sees excellent work in all these types. Most authors would probably aim at as wide a range of readers as they think may be interested, but if you aim too wide you waste weeks or months in getting a rejection: it is a serious breach of etiquette to submit the same article to more than one journal at once, and many editors require you to affirm that you have not done this.

Third, as long as the perverse misuse of IF ratings continues, most authors will make an assessment of the importance of what they have to write about and then aim as high on the IF rankings as they think they can succeed. An article published in *Nature* or in *Science* generates unusual kudos, though the reasons are not entirely clear. These journals are different in some ways from most of the rest. They look especially for novel concepts. Such articles are sometimes just a sketch for what will be published more fully later and elsewhere (Colquhoun 2007), but these journals do try to select accounts of work that is topical and will be of interest and significance to a wide range of scientists, not just to specialists. Both journals publish only an unusually small proportion (perhaps 10%) of the articles submitted to them. For that reason they are not usually likely to be sensible choices for a first solo effort. Seek (and take) advice if you intend to try one of these high-prestige journals.

Fourth are other features of the target journal that may be worth checking. Does it limit the length of an article, and will that be a problem for you? Does it make page-charges? Look for the gap between the 'Date accepted' notice that most journals add to each published article, and the date the journal part was published. Is that sort of time tolerable? Scan the Acknowledgements sections. Frequent recognition of help from, possibly unnamed, referees suggests that the journal uses constructive referees. Look at the figures. Does the journal format tend to over-reduce graphs so they are difficult to read?

Fifth, see if the journal is published by a learned society (science, but not primarily for profit) or is it owned by a commercial publisher (profit from science)? Some commercial publishers make excessive profits from science. In 1995 the average price of a scientific-society journal in the USA was 229 $ while that of a commercial one was 485 $ (*Nature* 419: 239.) Over the last two decades the average UK commercial journal price has increased by about four times as much as the UK Retail Price Index. In 2011 Elsevier made 37% profit (0.7 G£ profit on revenues of 2.1 G£; *The Economist* 14 April 2012). Some scientists are now refusing to referee for, or to submit articles to, these parasitic publishers.

There are good reasons (Moore 2010) for choosing a learned-society journal. Always consider one first in preference to a commercial journal.

It may happen that, after consideration, including discussion with colleagues, you are attracted to a particular journal but are still unsure that your work is within its scope. This should not happen often but if it does you can write a short letter to the editor (a 'pre-submission letter') including a summary of what you propose and asking advice about its suitability. Make clear that you understand that you do not seek a guarantee of acceptability, only that the subject and treatment seem to be within the journal's scope.

Whether or not to divide your work into what are sometimes facetiously called SPUs (Smallest Publishable Units, a process also called 'fragmented publication' and 'salami slicing') may be difficult to decide. Making five articles where one would do makes more work for you, more expense, and irritation for your readers, who have to seek each separate piece. It also wastes their and your time (and journal space) repeating parts at least of the Introduction, and probably the Methods too. On the other hand, university regulations for theses increasingly require several published or accepted articles ▼, and your chosen journal may impose page limits that are restrictive for what you need to say. And five publications on the CV, with different titles even though the contents are complementary or even similar, will look more impressive to busy people scanning rapidly but superficially. Similar pressures to atomise your work may arise from staff assessment procedures. In short, the public good requires that a single piece of work should be published as a single comprehensive article, but private need or page charges for a long article may incline you to several smaller ones, even though you recognise that such behaviour is reprehensible.

▼ A distinguished theoretical physicist describes (Dyson 2004) how, when young, he sought advice from an older colleague, X, famed for both experimental and theoretical skills, about work on a theory that he (the storyteller) had done with graduate students to explain some of X's experimental results. "I asked X whether he was not impressed by the agreement between our calculated numbers and his measured numbers. He replied 'How many arbitrary parameters did you use in your calculations?' ... I said 'Four.' He said '... von Neumann [a celebrated mathematician] used to say [that] with four parameters I can fit an elephant, and with five I can make him wiggle his trunk.' With that the conversation was over ... [I returned] to tell the bad news to the students. *Because it was important for the students to have their names on a published paper* [my italics], we ... wrote a long paper that was duly published ... with all our names on it. Then we dispersed to find other lines of work."

(Passage reprinted by permission from Macmillan Publishers Ltd on behalf of Cancer Research UK: *Nature* (2004) **427**: 297 *doi*:10.1038/427297a).

It may be difficult to decide what to do with collaborative work in which you have lost faith.

Finally, here are two warnings.

You may receive a letter or email soliciting an article for a journal you have never heard of. If it is not in the ISI database it may be a new, possibly commercial, journal trying to get established. Or it may be an established one with too few submissions, which should raise the question 'Why?' Be suspicious of such solicitations.

Lastly, you might have been invited, or successfully offered, to speak at a symposium or conference. The organisers may have told you that they intend to publish the contributions as a Symposium Volume or, increasingly often, as a Conference Proceedings or a Special Issue of a standard journal. (The editor of a primary journal is in a powerful position; the editor of a Conference Proceedings or a book composed of chapters written by different authors is more commonly a supplicant.) Or the organisers may not mention this until you have already agreed to speak. If the articles are not to be properly refereed then you have a strong reason for refusing to contribute anything written. The organisers start with high hopes of publishing rapidly (within a year of the conference). More often than not it takes longer, sometimes much longer than the organisers estimate (an example of the 'planning fallacy' – see the next section). A few years ago I was asked to submit a conference article about a small but (I thought) interesting piece of original work for a special issue of a reputable journal. Refereeing took 9 months. My revised version was with the guest editor for a further 18 months, and the article was published 32 months after it was first submitted.

Even if there is to be proper refereeing, consider carefully whether you wish to publish important original work in this way. At the start of the twentieth century duplicate publication was not uncommon, but it is now unacceptable except when an article is translated into another language. The derived version must be published only after the first; it must cite and refer to the original publication and to the fact that it is the second coming of this article; and you must get written permission for its republication from the publisher of both the first version and of the second. An author who is discovered to have achieved duplicate or near-duplicate publication covertly in the same language is liable to be blacklisted publicly (Malmer 1997), probably by both journals, with consequent damage to his or her reputation among peers.

In short: duplicate publication is strongly discouraged, so you may find that your work appears in a much delayed, obscure conference publication, with restricted circulation, so it does not get the attention it deserves, and prevents you publishing it in a more prominent place.

How long does it take to write and publish an article?

Research in cognitive psychology (and common experience) shows that on average it takes longer than you estimate to complete a task such as writing an article. This is known as the 'planning fallacy'. One of the reasons is the military adage: 'No plan survives contact with the enemy.' Nearly always a plan is blown off course by unanticipated events. The planning fallacy remains true even though you know about it and have experience in estimating.

Constructing tables and figures for a report of an uncontroversial piece of work that eventually occupies 10 printed pages may take one day to several weeks, depending on the complexity of analyses and your (un)familiarity with the computer applications you have to use. One has to assemble the data, perhaps make statistical analyses, discover how to get the word-processor to construct tables to your or the target journal's liking, and a graph-plotting program to make acceptable figures. Once you are familiar with the computer programs these processes become faster, but they are rarely completely straightforward. The higher your standards the more frustrating the processes and the longer they take. Writing the crude first draft of an article that eventually takes 10 printed pages may take only one to several days. But refining it, checking references, obtaining and reading those that you discover during the checking, re-revising, and the processes of internal review often need 5–10 weeks to pass, though perhaps only 1–2 weeks of actual work. And you probably have to fit all this in with other activities. By now 2 months have gone by since you first decided to produce the article.

As a rule of thumb, multiply your most realistic (hopeful?) estimate by 2.5 to give a likely time to completion.

Writing is, at least largely, under your own control; time to publication involves many other people and processes. You send your article to your target journal. Most journals aspire to rapid publication, and some achieve it, but they depend on rapid responses by the expert referees to whom the editor sends your article for comment, and are subject to inevitable delays in the production processes. There may be as little as 3 months between submission and publication, but a delay of 6–12 months is common. Many journals add to their articles the date of acceptance, and some add the date of submission too. From these and the journal publication date (supposing that to be accurate) you can assess the likely delay. If you need to be able to say that an article is published then start by assuming that it will take 6 months to a year (depending on the journal), and reduce that cautiously only if there is good reason to think it will be less.

Some advice

After choosing a target journal, and up to the first public draft

1. Print the requirements of your target journal: you will need to refer to these repeatedly. Most journals now require text to be word-processed, figures to be digital, and the article to be submitted to their website. Some may accept it as an attachment to an email. A diminishing few will still accept the article on paper sent by post.

   The commonest word-processors are (probably) those in the MS Windows 'Office' suites, in the Apple 'iWork' suite, and in the (free) Linux 'LibreOffice' suite, which is available for MS Windows and Apple too. All three allow you to set up styles that can be applied automatically to headings and paragraphs, thus freeing you from diversions into typesetting. Mathematical expressions are, unfortunately, not fully compatible among these three. The document creator 'LaTeX' has styles too, and is superior for mathematical expressions. It is a pity that many bioscience journals will not accept LaTeX documents. See what your target journal requires before you begin writing.

2. Next, construct all the tables and figures that you expect to need. They incorporate the evidence about which you are going to write. It helps to label the tables and figures with a unique tag that is at least partly non-numeric – 'Table A', 'Figure b', or perhaps 'Table T3', 'Figure F5' – and to refer to them in the text by these codes. Then if you add or remove one or more tables or figures you do not upset the existing labels. When the article is complete you can make global edits to create the necessary numbered sequences, for example 'Table A' changed throughout to 'Table 9'. This is a lot less

trouble, and much less error-prone, than trying to get the sequence correct during early drafts.

You can even save writing 'Table' and 'Figure' until the final global edit referring in the text, for example, to 'T3' and global editing 'T3' to 'Table 17'. You can also check before doing the global edit that your target journal does use 'Figure' and 'Table', and not 'Fig.' and 'Tab.'.

3. Many journals specify details such as font, type sizes, line spacing, page and line numbering. If yours does not, then set your word-processor to use 12-point type, double space lines, margins of at least 2.5 cm, and turn on page and line numbering (so referees can easily and accurately tell you what they are commenting on). Turn off automatic hyphenation, and set left-justification of text: the left margin of text is aligned but the right is ragged. (To justify both margins requires the addition of extra spaces between words and this can create unwanted effects in in-line equations and complex units, for example.)

4. Discover what your target journal does about citations, references, headings, and subheadings. For each level of heading in the hierarchy note the format (lower-case or capital, italic or roman font, centred or left margin), and the number of blank lines before and after the headings of different kinds. Then make suitable 'style templates' and construct all the main headings in the correct style (for example, centred capitals) before beginning to write. Many journals suppress the indentation of the first paragraph after a main, or any, heading. Style templates can deal with this automatically. A heading template may specify that the next template should be one that suppresses the indent on the first line, and that template in turn may specify that the next template to use should be one that includes the indent on the first line, and this template itself may specify that it should be followed by another of its own kind.

5. Compose the title page. Details differ among journals but all require a title. A few try to impose anonymity by segregating what follows on a separate page that is not sent to referees

('blind refereeing'), but most do not. Phrases such as '... in earlier work (Smith and Jones 1987) we showed...', and an abundance of references to the work of one person or group mean that in perhaps half of articles the name of at least one author could be guessed by the referees. A claim that double-blind refereeing was of advantage to female first authors was shown (Anonymous 2008) to be open to a different explanation, but double-blind refereeing costs little and may increase in popularity.

Next come the names of authors and their affiliations (addresses of the organisation that sheltered them while the work was done). Most also ask for a 'running headline' or 'running title' (that will eventually be printed at the top of alternate pages of the journal, though it is *not* required in that position on the typescript pages), and details of the 'author for correspondence' (name, address, email address, and telephone number of the author who is to liaise with the editor and usually with readers).

If an author's current address differs from the affiliation then it is usually shown in a footnote, as is the name and email address of the author chosen to respond to the editor and readers.

So far, so good. You have made a visible start. But many authors, experienced as well as apprentice, have difficulty in starting to write the body of an article.

6. Most people think much faster than they can type. One used to label a separate sheet of paper for each main section and jot down notes on the appropriate page. This still has the merit that you can carry the notes around and add to them wherever you are, independent of electronic machinery and batteries. But most people use a portable computer for this nowadays. Once you have the main headings then it may be helpful to dash down ideas, lists, arguments, citations, and everything else relevant beneath a heading but without worrying about grammar, style, details, or completeness. Next, go back and organise these notes into a logical order, amplifying as you go. Then rewrite the whole, concentrating on the manner in which ideas are expressed. These stages may be considered as three private drafts. By the time

they are complete you will be close to having a first public draft. What follows details this process.

7. If you are already clear in your mind about what is to go into the article, or have just completed the rough list of things to be woven into it, then you may decide to write section by section. The best order of sections for your reader might not be the easiest in which to write them, though. If you have difficulty then try writing the Methods section first. It is factual, probably straightforward, and requires little creative effort. Follow with Results and perhaps Discussion. (Some scientists prefer to start with the Results.) By that time you will be clearer about what must go into the Introduction and, probably last, the Summary or Abstract. These are suggestions only: there are no fixed rules. Do what works for you.

8. Once you have arranged your thoughts and notes for a section in a logical order, then write quickly without much attention to style or spelling, and leaving a trail of symbols or brief notes to yourself about details you need to check or amplify or modify later. I surround such temporary notes with '???', which is easily found later by the word-processor's 'Find' command. Word-processors also allow you to leave notes segregated outside the document margin. Keep the flow going at the same level of generality and abstraction, and avoid stopping and diverting from the main flow just to look up some minor detail.

Next, insert details you skipped over on first writing, and revise the section(s) for style.

9. If you habitually have difficulty in restarting to write, try stopping a short distance *before* a natural break in the composing process, and leave a marker there so you can jump to the place easily. (I use '###' for this.) Then when you come to restart you will know exactly what it is you have to write first.

Constructing the first draft has required an egocentric point of view: 'Look here at this wonderful bit of work that I did', though all the while remembering that you are trying to communicate with a reader. Now, however, you need to change point of view and look at the draft mainly from the point of view of an editor, a critical friend, a critical competitor, a reader.

10. When you have something close to a first public draft, shorten the line length temporarily by 3 cm or so (by changing a margin width), so that line breaks come in different places. It may be a surprise to discover how many errors or infelicities this reveals.
11. Check characters that are easy to confuse: O 0, l 1, and the roman and greek pairs aα, kκ, nη, pρ, uμ, *v*ν, wω, xχ. Some fonts (Times New Roman is one) make little or no distinction between 'one' and the letter 'l'. This is particularly serious if you want to use 'l' for the unit 'litre'. You may have to replace it by 'dm^3' (which is the formal SI unit anyway) or small capital 'ʟ'.
12. Emulate a potential reader: scan the Summary or Abstract, tables, and figures (with their captions) but without reading the rest of the text. This should convey the essence of your work and most of it should be intelligible in isolation.
13. Put this first public draft away for at least two days and preferably several weeks. Then take it out and reread it. How many things that seemed self-evident when you were writing have ceased to be so?

Most good writers revise several times.

Remember that there is a strong tendency to overcomplicate matters, and thus to obscure the main points of your work. Read what you have written, then prune ruthlessly. The less you write the more effective your article will be. A penultimate test of the King James Bible before final approval was to read it aloud. That may be a useful test when you think you have a completed article. Does it read well? Are you comfortable with each sentence and with the whole article? If not then revise – yet again.

At this stage, or earlier for multi-authored articles, you may need to send the draft article to other people for additions or amendments. It is easy to become confused about which file is the master and which contains proposed changes. Make it a rule to turn on the 'Record changes' option in the word-processor, and make sure that file names are always changed on return to you, and that the changed name includes the date the file was returned. Move all out-of-date files to a separate archive sub-folder, so the main folder contains only the current master copy with a unique name.

When several people are working on the same document they may be able to use one of the systems that allow all their changes to be put onto the same document. When they have finished then you have to read and accept, modify, or reject suggested changes in a single document. More usually however you will need to copy suggested changes from each colleague's document into your own (new) master document. Rather than asking for changes in parallel you may prefer to seek them serially, incorporating changes sequentially after each colleague returns their suggestions. This serial process takes longer, but later recipients see a version that is nearer and nearer to what all will have to accept finally.

Revision of the first public draft

14. Identify a colleague in your field, preferably a more senior one, whom you trust to be honestly critical. Graduate students often choose their supervisor for this rôle. Get the colleague to read the Summary and look at the tables and figures only. This is what a referee will probably do first. Can your chosen critic make sense of them in isolation? Then ask your critic to read the Results section. Can they now tell you what your work was about and what you discovered? The Results section should contain sufficient explanatory material to be understandable on its own. Listen silently to what your critic says, take notes, and do not waste time trying to explain what you meant or to justify what you have done. If your critic has difficulty then others will have too.

15. After that get the critic to read the whole article (if they will) and to point out, for example, ambiguities, misspellings, grammatical errors, illogicalities. The (biblical) beam in your own eye is more difficult to see than are the motes (specks of dust) in the eyes of others. If you are uneasy with English then at some stage get a native English speaker to amend the language. If the same person is your first reader then so much the better.

16. At this stage distribute the revised draft to a few others. The distribution list should contain any other authors, your supervisor

Hello Fresh

Say hello to the new way to cook!

A special voucher for **Amazon** customers

£40 OFF

CODE: PFG54K9R

"We're very excited about HelloFresh"
COSMOPOLITAN

"Genius!"
GRAZIA

Each week we deliver a box of recipes and fresh ingredients to your door

1 We create step-by-step recipes

2 We source exact fresh ingredients

3 We deliver everything to your door for free

4 You save time & cook great dinners at home

£20 off your 1st and 2nd box

How to claim your voucher

1. Go to HelloFresh.co.uk/specialgift
2. Choose your box
3. Enter **your code** at checkout

Code **PFG54K9R**

Valid for new customers only. Does not apply to gift and trial boxes. Upon redemption you will be signed up for a flexible subscription. Valid for UK and NI residents only, see website for full terms and conditions.

(if there is one who is not an author), and a few – two or three others – whose opinion you respect. This public review is probably the first time you get a chance to gauge the reactions of persons other than those already well informed about your work, and is usually an encouraging stage.

Submission to your target journal

17. Most journals now use a website as their preferred channel for article submission; some allow or use email attachment; a few will still accept submission on paper. Before sending the article off to your target journal add notes, using the word-processor's 'Add a note' ability (or handwritten on the paper if that is how you are submitting) to show where individual tables and figures are needed: 'Figure X near here' is a common form for such notes. Do not incorporate such notes in the main word-processed text: someone would have to strip them out individually before printing begins. The table numbers and figure numbers must appear in numerical order in these notes. Usually the table or figure should be placed near its first citation, but there are frequent exceptions where the first mention is incidental and the table or figure is placed later (suitably numbered) where it is considered in detail.

18. Reread the journal requirements. Does the journal require a signed declaration that the article has been submitted to it alone and will not be submitted to another journal while the editor is considering it? Does the journal require written confirmation from all authors that they are content that the article be submitted with their names as authors? Does the journal specify the electronic format ('.doc', '.pdf')? Does the journal require electronic versions of the figures? Some journals now put these and the text on a publicly accessible website as part of the pre-publication refereeing process.

Many journals are unclear about important details. What reduction will be applied to figures, and hence what size font should you use for the annotations on figures? How can you calculate the likely published length of your article (to avoid page charges)? Does the journal produce 'offprints' or '.pdfs' (*p*ortable *d*ocument *f*ormat) and, if so, when must these be ordered? (The usual answer is 'with the return of proofs'.) The paper offprint, that has replaced the earlier 'run-on' (mistakenly called 'reprint') technology, is still offered by some journals; the '.pdf' file is the electronic equivalent. Offprints are expensive while a '.pdf' is often free. Make sure that you will be allowed to copy the '.pdf' to enquirers, and check whether or not you will be allowed to make the '.pdf' freely available on your website: there may be copyright problems.

19. Compose a *short* letter to the editor formally submitting your article.
    - Say that you have tried to follow the journal's instructions to authors.
    - Say, even if not required to, that all the authors have agreed what is being submitted, that the article has not already been published, does not substantially duplicate another article, and has not been submitted to another journal. A single sentence like the one you have just read will show that you know what is expected.
    - Mention any closely related work that you have already submitted elsewhere.
    - You may also suggest two or three referees (do not name friends, or current or recent collaborators). Choose those who are knowledgeable in your field and who you think are constructively critical. The editor may be grateful, or may ignore your suggestions.
    - You should, if necessary, name persons who you think should *not* be asked to referee. If you do this, then give a brief explanation of your reasons. (Some journals have a space for this on their submission website.)
    - If you are aiming at a journal which you know is heavily oversubscribed with a consequent high rejection rate it is worth giving a brief explanation of why you think your article may be attractive to the journal. (Again, some journals have a space on their submission website for this.) The statement should look

more widely than the Summary can. Such journals have to operate an initial screening that is necessarily looking for reasons why they should not accept your article. Whoever does this screening will form an opinion and be biassed by the quality of the submission letter and the Summary (and perhaps the figures). They may even make their judgement on these alone. You have to get over this first hurdle. For this reason if for no other the submission letter and Summary need care.

20. Make your own copies of what you are about to submit, and read carefully again the journal's instructions about what and how to submit. Then press the button (or mail the packet)!

You should receive a prompt acknowledgement that your article has arrived. If you hear nothing within a week then enquire by email to the editor.

The editor and the publishing process

Your article may be judged by the editor in person or by a member of the Editorial Board, or by a professional assistant editor if the journal is a big general-purpose one such as *Nature*, *Science*, or the *British Medical Journal*. But if your article survives initial scrutiny the editor will probably send it to two or three referees to get their advice. (Referees are often called reviewers, particularly in the USA, but I use referee to avoid confusion with the authors of those articles that review and summarise a whole field.) Referees try to assess the originality, significance, and competence of your work, and the clarity of your presentation. They are asked to report within a few weeks. Their first duty is to the editor and thus indirectly to the journal readers ▼. Some referees seem to feel that they will acquire merit in proportion to the extent and savagery of their criticism but a good referee will be constructive and will also try to help you to avoid making a public and permanent display of your incompetence. The referees' comments are likely to be made as one or more of: (1) most commonly nowadays, comments or notes inserted electronically into a copy of your work; (2) comments in a separate file, using page and line numbers in a copy of your article as cues; (3) rarely nowadays the older and more satisfactory method of annotations on a paper copy of your work in pencil (ink, particularly red ink, is aggressive); or on separate sheets of paper.

> ▾ Refereeing is a voluntary, usually unpaid, and important part of the whole scientific process. Editors are always seeking scientists who are establishing a reputation. Being asked to referee is a sign that you have been noticed. If you are diligent and helpful the invitations will pursue you to the grave, and beyond. Good referees are often amongst the busiest people, so the ideal of a reply in 2–4 weeks can in practice become 1–2 months.
>
> Henry Oldenburg, the first editor of the *Philosophical Transactions of the Royal Society*, seems to have begun the practice in the 1660s, seeking expert opinions on articles submitted for publication. The use of referees increased erratically. It has been usual since the 1940s and is often called 'peer review'. "At its best, it provides prompt, detailed, constructive and well-founded criticism to the benefit of researchers and consumers of research. At its worst, it is expensive, slow, subjective and biased, open to abuse, patchy at detecting important methodological defects, and almost useless at detecting fraud or misconduct." (Passage reproduced from Godlee F, Jefferson T, Eds., 1999 *Peer Review in Health Sciences*. London, BMJ Books.) Perhaps the biggest danger is that an article by author A will be sent by the editor to referee B, the best-informed person in the field, who may also by that same fact be a direct competitor of A for acclaim and grants (Broad & Wade 1982). In spite of these dangers the system is usually at least satisfactorily effective, and no acceptable substitute has yet been devised. Peer review by referees is primarily to advise the editor. It seeks to improve articles: it cannot validate the work they contain.
>
> Peer review of applications for grants raises some similar and some rather different questions (such as looting of original ideas). It is outside the scope of this book.

> Articles typed on paper used to reach a referee by post (mail). Comments, large and small, could be written as annotations on the paper. Minor amendments to wording were easily suggested *in situ*. Most journals now send articles to referees in electronic form and expect referees to comment electronically too. On a word-processed '.doc' file comments can be made using the word-processor's note system, but minor amendments are more tedious than on paper. Some journals send a '.pdf' file of the article on which it has been made possible to make minor amendments, but not major ones, unusual ones (such as the correction of Greek letters or sub- and superscripts). The software is improving though. Some, mainly older, referees still print the article on paper and annotate that. Such a referee bears the cost of printing and posting the paper copy back to the editor. Either an easy electronic substitute for annotating will emerge or detailed amendments will become rare – a noticeable loss to authors, readers, and editors.
>
> The intended savings in time using email are not great, and are usually less, even much less, than the time the article lies on the referee's desk waiting for attention.
>
> That electronic refereeing is possible cannot be doubted; that it improves things for authors, referees, readers, and even editors, is still debatable.

Referees should treat your article as a confidential, legally privileged, communication. Until it has been published they should neither communicate it to, nor discuss it with, anyone. Nor should they make use of any of the ideas it contains for their own or others' benefit. It has to be said that some journals and referees are not as strict about these matters as they should be, but serious abuse seems to be less common than the suspicion of it.

In the recent past most referees were anonymous, but there is a trend for editors to ask, or at least to allow, referees to identify themselves to the author. Identified referees may feel more strongly the need to be careful and reasoned in what they write, and to phrase their comments temperately. Against that may be a reluctance to be identified as the honest writer about defects in a friend's work, or downright fear of being known to be a critic of the work of a senior or influential person in the field.

A good editor will choose constructive referees, and will evaluate their reports, then compose his or her own report to you. Often nowadays, however, you get simply a copy of the referees' reports and the editor's decision. The decision is usually in one of four categories.

1. Accept (probably with some minor revision).
2. Revise (with the unstated hint that if this is done to the editor's, and perhaps the referees', satisfaction the article will be accepted). There is a tendency to replace 'revise' by 'reject' but with a form of words that allows resubmission after rewriting. The published submission date will then be that of the rewrite, thus improving the journal's apparent speed of publication.
3. Make a major rewrite (and no hint of likely acceptance, but at least not rejected).
4. Reject because ... (unsuited to the journal, too trivial, not interesting enough, incompetent – though more politely expressed, of course).

When you receive your article back from the editor, take a deep breath and skip rapidly through the editor's and referees' comments. There will almost certainly be criticisms. Are they matters of substance, or just minor matters of presentation? Then read again more carefully. Perhaps the referees have misunderstood something you thought was obvious. That shows that it was not as clear as you supposed. Perhaps they are encouraging you to reorganise and extend the Discussion? Perhaps they are right to do so? They may be asking for more work. Is that reasonable? Is it practicable? Note by each point what you can do about it.

If the editor's decision seems to allow resubmission and you decide to do that then go through the list attending to the minor requirements first and then the more difficult ones. Give this work fairly high priority: no editor likes to have articles that have gone into limbo, and many set time limits of a month or two for a resubmission.

Whatever else you do, do not be abusive when you reply to the editor. The referees – busy people like you and your prospective readers – have probably tried to be helpful. Accept their justifiable comments with gratitude as a part of the publishing process, and respond coolly, logically, and with a good grace where you think the referees or editor were actually wrong or ill-advised about something that matters to you. Most published articles have been improved by the referees' suggestions, some out of all recognition.

Include with the revised article an itemised reply to the comments and suggestions made by referees and editor, explaining briefly what you have done about each (or why you have done nothing). Some editors ask for such a list, but you should include one even if it has not been requested explicitly.

What happens to your revised article depends on the editor. A good editor will have used the referees' original opinions for guidance and will now read your revised version carefully and make his or her decision. Some editors, however, are mere clerks and behave like the Duke of Plaza Toro in Gilbert and Sullivan's opera *The Gondoliers*, who "... led his regiment from behind: he found it less exciting". They may not read your revised article carefully, and simply send it out to the original referees, whose opinions they then accept. Most referees are unenthusiastic about this, believing that the editor should now make his or her own decision, so a second review is likely to be cursory at best.

If your article is rejected, recall first that almost all established scientists, even celebrated ones, have had articles rejected, some being told that work that later became famous was unimportant or uninteresting or insufficiently complete. Examples are the biochemist Hans Krebs (the Krebs cycle), the physicist Enrico Fermi (beta decay), and the theoretical physicist Peter Higgs (one of the predictors of the Higgs boson). If you think the editor's decision was wrong then you may be able to alter his or her mind by reasoned rebuttal and argument, but if that fails then you can take the article to another journal. The city of Science has many gates, and if one of the gatekeepers refuses you entry, you can always try another gate. This is one of the strengths of scientific publishing. If you do decide to try a different gate it is sensible to use the referees' comments from your first attempt to revise your article, and you will usually need to modify it to fit the detailed format of your new target. The editor of your new target journal is quite likely to send your article to one or more of the same referees as the original editor did. Such a referee ought to, and probably will, refuse to act a second time.

When your article is accepted it will be copyedited for details to make it conform to the journal format (and perhaps in minor ways to the editor's views about grammar and syntax). A short time later you receive 'proofs'. These used to be on paper, but nowadays are more likely to be electronic. You are asked to return the proofs with essential corrections either yesterday or sooner. (Proof correcting is described later.) Corrections at this stage may be expensive, and the editor may threaten to require payment if you ask for an excessive number that are your fault. You may be irritated at the delay since acceptance combined with the demand that the proofs be returned quickly and the threat of charges for undefined 'excessive' changes. Suppress your annoyance. Your article has been through several processes, of which proof-reading is nearly the last. The threat of charges is to restrict changes to essential ones.

Electronic publishing of individual articles as soon as they are ready is eroding the idea of a 'part' or 'issue' of a journal. Articles published singly in this way may be cited by their document identification number (*doi*: Chapter 11) rather than the conventional journal name, volume number, and pages. Some journals publish articles individually online, and then subsequently assemble them into a part or volume. An article starts with a *doi* and acquires a conventional reference later, and can then be cited in either or both forms.

For those journals that still publish only on paper and in 'issues' or 'parts' containing several articles, it is a cardinal principle, sometimes ignored, that no article should be published until the proofs of all the articles in the 'part' have been agreed by the authors, or by at least one author with power to act for any others. To 'pull' (delete) an article from an issue of a journal because the proofs have not been returned is expensive and error-prone because Table of Contents, indexes, and all the page numbers following the deleted article have to be changed and so do all references to them. Even with electronic publishing programs the editor and publisher try to avoid this process. So when you are about to go away for several months, or will be unable to use email, do make arrangements for a co-author or trusted colleague to deal with proofs. Many journals' websites allow you to change the corresponding author – the one responsible for your article, and particularly for dealing with the proofs.

The editor's duties

What are a good editor's duties? They are three.

First, the editor is responsible for attracting, or soliciting, and selecting accounts of work that is of high scientific value and interest. A journal that succeeds in this will prosper; one that does not will falter or fail. The pernicious influence of impact factors – league tables for journals – has sharply increased the visibility and perceived importance of this duty.

Second, the editor has a compound duty to authors and to readers: to help authors to express themselves as clearly and concisely as possible, so that the readers' task is made as simple as possible.

Third, the editor is responsible for ensuring that the administration of the journal is accurate, ethically acceptable, humane, and as rapid as possible, and that the production standards are as high as possible within the constraints set by money, time, and people.

In practice you may find a wide range of approximations to this ideal.

Proof correcting

There used to be two main sorts of proof on paper: 'galley proof' (in long unpaginated strips) and 'page proof' (pages, with temporary or no numbers). This two-stage process was needed because the text had been 'set' by a typesetter (compositor) who was unfamiliar with the work, and made mistakes. Nowadays you have yourself set the text so there are (or at least should be) many fewer mistakes. Science journals therefore skip the galley proof stage and provide only page proofs. Final page numbers and cross-references to them are added by the journal. Until a decade or two ago you got proofs printed on paper, and were able to write corrections on the proofs. Nowadays you get some sort of electronic version to correct. Electronic proofs may save distribution costs and time, and they save the publisher money. But they are less easy to correct than paper proofs and the process is often frustrating, error-prone, and time-consuming for the author. The commonest proof is a '.pdf' file that, unless you have an expensive application program, allows limited changes: insertion, and replacement or deletion of text selected with the mouse. It may be difficult or impossible to select what you want alone, and, for example, Greek characters, sub- and superscripts, accented characters, italic face may be unavailable or difficult to import. Irregularities in formatting may be impossible for you to adjust. We are still at the Henry Ford stage of "You can have any color you like as long as it's black" (because black paint was the only colour that would dry fast enough for the assembly line speed). But the software is improving, and these restrictions may soon be less serious. If the corrections are few and simple the '.pdf' proof may be adequate. When more complicated amendments are necessary, some authors reply with an email containing a detailed description of necessary corrections or changes. If the amendments are difficult to describe unambiguously this may not be adequate. You may then prefer to use the older way: print the electronic proof on paper and then mark that in the traditional way and post it to the journal. The next two paragraphs refer to this sort of paper proof. What follows them applies to electronic proofs too.

There are standard ways of indicating changes in proofs on paper. Each change requires a mark and perhaps some text in the margin and a corresponding (but usually different) mark in the text to show exactly where the change is needed. There is no universal standard set of marks but most sets are similar, and the meaning of most marks is obvious. The commonly needed marks in one set are shown in Table 1.3, and in BS 5261-2 2005, the British Standard of proof-correction marks (British Standards Institution 2005).

The colour you use to show changes may matter. By convention green is reserved for the professional proof-reader (unlikely nowadays). Your journal may have its own conventions, but if not then mark in red any errors introduced by the copy-editor or printer after the article was accepted ('not *my* fault') and in blue or black the changes that you have decided must be made ('my fault').

Superficial blemishes – 'morality' where you had intended 'mortality', 'relative humility' for 'relative humidity', 'conversation' instead of 'conservation' – may be obvious, but that your offspring has six toes may easily escape your notice. Pride in this first sight of the printed article may dim the critical faculties of the most vigilant of authors. *Beware!* Try to imagine that the article is written by someone else you know to be careless, and assume that it is full of errors: that references to figures in the text have been corrupted, that numerical values in tables have been transposed, that a block of references has disappeared, that your name has been misspelled in the running head at the top of (alternate) pages, that the wrong level of heading or an inaccurate cross-reference has been inserted. Even though your electronic file is the source it has been extensively tinkered with by at least one, well-meaning but probably harassed, copy-editor.

Read through the proofs several times *looking for only one sort of thing at a time*. Make provisional corrections on a working *copy* of the proof.

- Check the title, authors' names, and affiliations.
- Check the running head on *every* page.
- Check all the section heading formats.

TABLE 1.3 Proof-correcting marks

Mark in margin	Meaning	Mark in text	Mark in margin	Meaning	Mark in text
/	End correction in margin		Greek ital. alpha	Change as specified	$y = \alpha p + 1$
⟨#	Insert space	for⟨an example	move	Move to position shown	While were the case showing
⟨ an#	Insert new matter	for⟨example	n.p.	Begin a new paragraph	for an example. If this then
⟨ ⊙	Insert punctuation	for an example⟨ When	run on	Run on as one paragraph	for example. But if this
⟨ an	Substitute	for my example	more #	More space between lines	the end. Then open up
⌒	Close up	for an ex ample	less #	Less space between lines	The end. Then close up
⟨	Delete matter struck through	for any example	over	Take over to next line	rate of 5.1 m s⁻¹ is small
⟨	Delete and close up	for an exampple	back	Take back to previous line	rate of 5.1 m s⁻¹ is small
trs.	Transpose	an example for	(EDITOR)	Note to editor	
stet	Leave as printed	for an example	(PRINTER)	Note to printer	
ital.	Change to italic face	Be **CAREFUL**	⌐	Move right (to align)	to the end. Then close up. If not, then
rom.	Change to roman face	(Be)CAREFUL			
caps.	Make uppercase	Be *careful*	⌐	Move left (to align)	to the end. Then close up. If not, then
l.c.	Make lowercase	BE CAREFUL			
bold	Make bold	Be CAREFUL	∪	Superscript	x(3)
underline	Underline	Be **CAREFUL**	⌒	Subscript	a(n,t)

- Check the sequence of table numbers.
- Check the sequence of figure numbers.
- Are the tables and figures in sensible (or at least acceptable) places?
- Have the figures been reproduced at an acceptable size? Are they easily readable?

- Are the contents of tables correct? They should be what you submitted but mysterious changes do happen. Anyone can check words, but only you can check numerical values.
- Check the citation and reference details, and complete any references that have been printed since your article was accepted.
- Check that no references have been omitted or given a date that conflicts with a citation.

Begin reading the text only after you have checked all these things. Professional proof-readers would check each letter against the copy-edited text – they might even check everything backwards to avoid being influenced by the sense. You, however, should be more concerned with the sense of what has finally appeared on the page. The copy-editor may have made changes some of which you welcome and some which you do not agree with. You will have to accept those changes in spelling made to conform with the journal's version of English, but you do not have to accept all the journal's fetishes if you have a good reason for rejecting such things as (for example) split infinitive abhorrence, or revulsion at starting a sentence with 'And' or 'But'. If the change is trivial and does not alter the sense then it may be necessary to let the change stand though thinking 'I wish they hadn't done that.' (Every change requested adds to the risk of introducing a further error.) If a copy-editor has changed the sense of something you wrote then you must correct it, but look carefully to try to understand why the mistake happened. Perhaps what you had written was ambiguous? If possible design corrections and changes to occupy the same space as the matter they replace, so that the number of lines and pages is unaltered.

Finally, if possible, get a colleague who is not an author of the article to read the proofs and look for errors that you have missed. The vision of a good friend will be less clouded by emotion than yours is.

Journals that still produce 'offprints' usually ask that you return an order for them (if you want them) with the marked proof. Or you may be offered a free '.pdf' file of the finished article. They may also ask at this point, if they have not already done this, for a declaration assigning copyright to the publisher.

That is the end of the mechanics of writing an article and getting it published. If you are anxious to get started on your own article then now is a good time to do so. But it is worth reading quickly what follows about copyright, and about the current revolution in publishing. These sections complete Chapter 1.

Intellectual property: copyright and patents

Intellectual property is a usefully vague term that covers a variety of legal rights that give protection to, for example, writers, artists, scientists, inventors, designers, and investors. Examples include copyright, patents, trademarks, plant variety types, design rights, database rights, moral rights, and computer algorithms. Copyright law is important to all authors; patent law may be to some. Both give effect to the belief that it is morally wrong to plagiarise. The State grants for both a partial monopoly over use of the protected material for a specific limited period in return for the copyright or patent holder eventually allowing the work or invention to enter the public domain where it will be free for anyone to use.

What follows gives you the background to the dealings you are likely to have with copyright.

In most countries, copyright arises automatically – you do not have to apply for it and need not register it. It applies as soon as an original idea has been 'fixed' in some non-ephemeral tangible form (known as 'material form') such as written or printed words, illustrations, maps, photographs, or magnetic, optical, or electronic recordings (all of which are known as 'works'). It is easy to imagine disputes about what constitutes 'material form': writing on snow? with smoke in the sky? on water with a jet ski? in electronic RAM (Random Access Memory: where the work disappears when the power supply is switched off)? face-painting? What is 'substantial'? What is 'non-ephemeral'? And what are the bounds of an 'original' work? A train timetable? a telephone directory?

databases in general? Is there copyright in the design of an integrated circuit (yes) and in a computer program (yes, but only in some circumstances and places). These are the sorts of question that have gone to the courts. The idea underlying copyright is that it protects expression but not ideas, which are free for everyone to use. Copyright protects only the particular material form, such as a book or CD or e-book, by which an idea is communicated. The reasoning is that this will encourage people to create new 'works' that will eventually pass into the public domain, thus increasing the total amount of knowledge in society.

A patent, in contrast, must be applied for, and the Patent Office will grant protection only over a 'novel' invention that displays an 'inventive step'. In effect, patents protect original ideas that can be registered because they disclose technical information. Getting a patent is an expensive and slow process. To protect the novelty of your invention, you must avoid 'prior publication' of the idea you wish to patent until the patent has been at least provisionally granted. This requirement is likely to conflict with the generally communicative ethos of scientists.

Each country has its own laws governing copyright and patents, though these are increasingly convergent. The period of protection for copyright differs among countries and is changed frequently, usually increasing. Lifetime of the author plus 70 years is typical at present. Patent protection generally lasts 25 years ▼.

▼ English copyright law emerged after the invention of printing with moveable type. The Stationers' Company (a livery company) whose members combined the rôles of printer, publisher, and bookseller, persuaded Henry VIII to grant them a monopoly and to ban the import of printed books. By 1662, the Licensing Act had established a register of licensed books and required the deposit of a copy of all such volumes. In 1710 the first copyright law, known as the 'Statute of Anne', was passed, ostensibly "for the Encouragement of Learned Men to Compose and Write Useful Books" but probably really to protect the monopoly of its promoters, the Stationers' Company, while allowing the Crown to censor and effectively prevent the publication of seditious, revolutionary, texts. It gave protection for 14 years with one renewal for a further 14 years, and these short periods may indeed have spurred authors to continued writing. Later extensions to more than the lifetime of the author must have different purposes (perhaps because many copyright owners are companies with potentially unlimited lives). This first law also allowed appeal to the Archbishop of Canterbury against misuse of the monopoly – a provision that was soon dropped. The first USA copyright law was passed 80 years later in 1790 and protected only American authors, a situation which changed little throughout the following century. The international Berne Convention of 1886 brought about 'national treatment' laws, which required signatories to treat other member states' authors as favourably as their own. The USA did not accede to the Berne Convention until 1989, but then faced about and argued strongly for robust, uniform standards of copyright protection throughout the world.

Many countries allow you, for your own private study or research, to make a single copy of part of some types of copyright work (such as literary, dramatic, artistic and musical works, but not broadcasts, sound recordings, or films) without violating the author's copyright. This is 'fair use' or 'fair dealing'. It can also be fair dealing for the purposes of criticism or review to quote part of a work. Universities and similar institutions often have group subscriptions allowing production of single copies of journal articles. There is no simple rule about how much a work constitutes 'fair' dealing: take expert local advice. Copying three pages from a book of 100 or more pages is probably acceptable; one chapter may be; the whole book except for a few pages is not.

Ownership of copyright and patents is governed by specific laws but also by contracts. Unless there is a specific agreement to the contrary (such as a clause in an employment contract or license agreement) several presumptions are likely to apply. For example, if you work in a UK institution that employs a full-time specialist photographer or illustrator then you may expect that copyright in their work that was made in the course of their employment will belong to the institution. But if they were hired and paid as self-employed persons to produce a particular work, and are not on the payroll as full-time or part-time employees, then, unless contracts state otherwise, you may expect that copyright belongs to the specialist, *not* to the institution, even though the institution paid for the work to be made. Such a payment is usually for the right to reproduce the work in a single context – a single journal article, for example. Copyright stays with the person who made the 'work', and further reproduction requires his or her permission. These arrangements are not intuitive and lie at the root of many misunderstandings and disputes. There may be doubt too about the ownership of copyright in an article you have written about work done in the course of your employment. Unlike the photographer, you were not told to do a particular piece of work, only that work of this generic kind was expected. In the past, most institutions have accepted that copyright in such works belongs to you, the author, but some institutions are inclined to dispute this.

The laws of copyright evolved to deal with the concept of 'literary author', but scientific 'works' with numerous 'authors' have diverged a long way from this simple situation. The law recognises as authors only those persons who actually contributed written words or images to the article. If you describe a method used by a specialist who contributed to the work, and the specialist simply agrees that your description is accurate, then the inclusion of the specialist's name in the list of 'authors' probably does not confer legal authorship on the specialist.

Your first direct contact with copyright may be when the editor, having accepted your article for publication, asks you to assign your copyright to the journal publishers. You probably did not intend to make money directly from writing the article so the implications of assigning copyright may seem trivial, and most scientists assign and forget. Pause a moment, however. If you are on the payroll of an institution and your article was written in, and about work done in, the course of your employment then the copyright should perhaps be assigned on behalf of your institution (this will depend on whether your employment contract or some other agreement with your employer has indicated who owns the copyright in the work you produce). Some institutions understand this and make suitable arrangements. Many do not. And if your article contains illustrations or photographs commissioned from a specialist who is not on the payroll of your institution then neither you nor your institution can validly assign that specialist's copyright.

A commercial publisher may require payment for giving the right to reproduce commissioned copyright material in a specialist scientific book or review, but rarely for material from a primary scientific journal. It is ironic, however, that commercial 'for profit' publishers may sell the right to enter a work on a commercial database. The database owner may then charge for access to the database, including access by the institution that sheltered the work in the first place.

You are most likely to see the other side of the copyright fence if you write a review, or a chapter in a book, or a whole book, and want to reproduce images or long passages from the work of others. Your publisher will probably require you to get permission for each instance. Most publishers now deal with such requests on their website, and this greatly reduces the administrative burden. But courtesy, at least, still requires that you get permission from the author too if you can.

The digital electronic future

We are in the midst of a publishing revolution – the biggest since the development of printing with movable type five centuries ago – as results of the explosion in digital electronics and of the invention of the internet ▾. Three features of this revolution are important.

> ▾ The essence of the internet is that control of routing is distributed, so it is possible for parts of the system to fail but the rest to continue autonomously. It developed from an American military project, ARPANET (Advanced Research Projects Agency NETwork), undertaken in the shadow of nuclear war. Its successor, the internet, is now near ubiquitous. It allows quick access (in principle) to most published articles, though financial restrictions limit access in practice.

First is the mechanics of the publishing process. Most journals now require authors themselves to prepare their article (text, tables, and figures) in electronic form, and to submit it over the internet. Mistakes during composition of the article can be much more easily corrected than they used to be on a typescript, far fewer professional typesetters are needed, and what is submitted should contain fewer mistakes.

Second, storage is vastly cheaper than it was when all journals were published on paper only. A USB 'stick' can hold more than 10 000 100-page paperback novels. Many journals now allow you to put data and details (text, tables, figures, short films, sounds) that supplement the published article on a database that they maintain and which can be viewed by anyone on the internet (and, if required, with a subscription to the journal).

Third, these and other changes encourage publishers to experiment. For example, in 1999 the *British Medical Journal* began to report work at full length in electronic form on a website and at the same time more briefly in print. The full version is for fellow scientists in the field who need to see the technical detail; the short form is for those in a wider field whose chief need is for the results and discussion. Many journals now publish both online and paper versions of the same article.

Some journals publish online only – less costly than printing and distributing on paper.

Three decades ago, some journals began to require the author to pay page charges, at least for what the journal considered to be excessive length. In general, however, the structure of scientific publishing, sanctified by the centuries, loosened but changed little: an author did not pay to have an article published, but individuals and libraries paid to receive the journal. This is now known as the 'subscription' or 'reader pays' or 'restricted access' system. Few institutions could afford to subscribe to all the journals that their employees needed, or wanted, to read. You learned that an article had been published by reading a *Current Contents* or similar list, and eventually you could read the abstract in, for example, *Biological Abstracts*. To get the full text of an article in a journal your institution did not subscribe to you could ask the author for a 'reprint', or go to a library that did subscribe to the journal. Improvements in photocopying eased things a bit, sometimes illegally, but in essence the process was still closer to the horse-and-cart than to the internal combustion age.

As digital opportunities developed, several questions emerged. Here are a few.

1. How would standards be maintained? Some journals experimented with 'open refereeing' in which a submitted article was made openly available on the internet for anyone to comment. Others tried post-publication modifications based on criticism after publication. The problems with such schemes are easy to imagine.
2. How permanent would the 'published' article be? The fundamental form had become a digital file rather than print on paper. This file might be held on a magnetic or optical medium. One might think that this has an indefinite life, but that is not so. Magnetic media, particularly tapes, deteriorate over even a few years and so do optical media. Even more rapid at present are the changes in physical form and in formatting – layout on the medium – so that it becomes difficult or impossible after a few decades to find

apparatus that can read the medium. Are 1" magnetic tape reader/writers still used much? Could I recover data from a 1990s 15" hard disk or 8" floppy? When, if ever, did you last see a working paper-tape punch and reader or a card punch and reader? They were common in the 1980s. File formats change too. Who will ensure that all the accumulated files are transferred to the new medium and perhaps a new format every time that the old one becomes obsolete?

3. The very idea of the 'part' of a journal began to give way to almost continuous publication of individual articles as they became ready. In that case, would one subscribe to a journal at all, or just pay for the right to read or to copy individual articles? Most journal websites already let you see a summary of an article but require payment to see the full article.

At this point a powerful side current thrust itself into the hitherto smooth flow, creating violently choppy water. What follows here is a mere incomplete (2013) sketch of the main features of the turbulent and ever-changing scene. The sketch is illustrated by the position in the UK, though many of the problems and consequences apply elsewhere. Worlock (2013) gives a more detailed account.

Grant awarders (such as UK Research Councils, and charities such as the Wellcome Trust), who pay for about 60% of the work that results in a published scientific article, began to advocate, and later to insist, that all scientific articles resulting from work they had paid for should be available, without charge, on the internet – the 'open access' (OA) model. Work they had paid for thus has maximum accessibility. It requires that there be no more than minor copyright limitations and that the author consents. The OA principle now has wide support. A new generation of online-only not-for-profit journals began OA publishing, charging fees to the author to cover their costs. An example is the *PLoS* (*Public Library of Science*) series. In 2011 *PloS One* published almost 14 000 articles. A recent (2013) addition to the OA field is *eLife*, devoted to the biosciences, and backed by the

Wellcome Trust, the Howard Hughes Medical Institute, and the Max Planck Society.

The key procedural change here is: who pays? The costs of website only publication might in principle be smaller than those of publication on paper (Delamothe & Smith 2004), though this is not as clear in practice. Commercial journals want to protect their operations, including their profits ▾. Many learned societies get a large part of their income, which they spend on their education and subject support activities, from their journal subscriptions. So research councils, charities, journal editors and commercial owners have been in intense discussions for several years, and new arrangements are still emerging.

> ▾ WARNING. Over 8000 journals calling themselves open access have erupted in recent years from the murkier depths of commercial publishing (including some well-known names). They charge to publish, and claim to use referees, but Bohannon (2013) showed that over half a sample of 304 of the more-credible such journals accepted and published a spoof article that contained such severe and obvious errors that it should have been rejected out of hand.

The influential *Finch Report* juggled this very hot potato. Several schemes for OA of publicly funded research emerged, named with colours. Two have become preferred: 'gold', and 'green'.

For gold OA the author pays the journal to publish the article which is immediately available unrestricted on the publisher's website. *PloS* charges about 1200 to 2400 £/€/$ for gold access; Elsevier charges 450 to 4500 £/€/$. Both charges depend on which of their journals you choose.

Green OA needs no payment by the author (except perhaps page charges for long articles), but the article may be made available without restriction only after a defined period. If the journal offers gold OA but the author lacks funds for this then green OA is allowed after 12–24 months. If the journal does not offer gold OA then green OA is available after 6 months. After that time the author is free to deposit the article in a publicly accessible archive. Some archives are run by institutions. Others are run for a community, for example, *PubMed* (for biology, especially medicine), and *arXiv* (run since 1991 mainly for the physics community but now accepting some biological articles). These archives are repositories only: they may accept work that has not been published anywhere else, as well as fully edited and published articles, but they do no detailed peer reviewing.

The *Finch Report* preferred gold OA, but that raises the question: where is the author to get the money? This is a new and, for many, a worrying question. UK Research Councils and charities will no longer grant money for publication charges, as they used to do. The UK Research Councils have transferred some money to UK universities for this purpose, but the amount is much less than is needed. About 40% of published articles have no grant support anyway. Many journals seem willing to consider waiving, or at least mitigating, their charges to those with no overt external support. Some institutions seem willing to consider supporting publication by their members, but they may restrict it to high prestige journals. In general the search for funds for publication is a new area of responsibility for authors. Let us hope that fears may prove liars.

What next? It seems likely that those journals that publish on paper (and possibly online as well) will continue to do so for the time being at least. Elsevier has said that under this hybrid scheme it will adjust its journal subscriptions to take account of income received for gold OA. There may be a drift towards the *PLoS* and *eLife* model of internet-only, OA, author pays, publishing. As OA becomes commoner then libraries will become less willing to subscribe to paper journals. Perhaps savings there may be used to support gold OA to journals. The transition from the old stable state to a new one is likely to take a decade or more, with constant changes and confusion along the way.

Amid this ferment there are a few fixed points. Scientists are always too busy to read all that they would like (or ought) to. So there will be continuing pressure to keep articles as short and clear as possible. The general principles in the preceding sections and in Part II are likely to remain as relevant as they are now.

Interesting times.

2

Speaking about your work

You will probably get an opportunity to talk about your work before you have to write about it. You use some of the same evidence in a talk as in an article, though prepared differently, mostly in visual form with tables and figures as 'slides'. But speaking is a performance art in real time and needs different skills from writing. When writing you have time to reconsider and revise; when giving a talk you have only one chance to get it right. Detailed preparation and at least some practice are essential. Giving a successful talk that interests the audience can be a satisfying, if nerve jangling, experience. It is a rapid way to recognition amongst peers in your subject ▾.

> ▾ The distinguished mathematician Richard Hamming, in a remarkable lecture (Hamming 1986) about how to do really important research in science, as he himself had done, concluded that giving a polished talk about the work was an important part of the process. The whole text of his lecture is thought provoking. Do read it.

When writing or showing a poster you are competing for attention, but a talk is different. Your audience is captive. This advantage is also a responsibility: the members of your audience have come hoping to learn something and, perhaps, be entertained too. The stakes are high.

The main aim in giving a talk is the same as in writing:

to communicate with the members of your audience as clearly and simply as possible.

There are usually six phases to giving a talk: writing a summary; preparing what to say; making visual aids; practising and refining; preparing on the day but before your talk; and the performance itself.

(1) Writing a summary

The meeting organisers will probably ask for a summary long before the meeting. This summary will be printed in the programme of the meeting. The length may be limited to anything between 50 words (and no illustrations) to one or two A4 pages with one or more illustration(s). This is the time to give serious thought to what you want to communicate: too often a summary of a planned talk is written in anticipation of results that do not materialise. The questions before writing a summary are similar to those you ask when composing the Summary in a written article: what audience are you to speak to? What do you want its members to remember? Ideally, the contents of the summary are similar too: concentrate on what you wanted to find out, what you found, and its implications, and be as numerically specific as you can. But talks are often (usually?) about current work the details of which are simply not available far enough in advance to be included in a summary. It may be better, therefore, to make the title and the contents of the summary as general as possible. Unsatisfactory but difficult to avoid.

(2) Preparing what to say

As in writing an article, the first thing to do is to decide who and how large your audience will be and to decide how to view its members. Is it a few dozen fellow scientists in a subsection of a learned society (self-selected specialists)? Or is it all your colleagues in the department (scientifically literate, open-minded, but mostly only vaguely aware of your subject)? Or, again, is it a public talk to the Women's Institute (lively and intelligent but mostly non-scientists)? You know your own subject inside out and back to front. It is horribly easy to assume, mistakenly, that the members of your audience know more than they actually do, and to try by over-complicating to anticipate expert criticism. Readers of a written article can go back and reread, but the members of the audience at your talk cannot. They then hear and see what you present but fail to understand, so they switch off. Your message may survive a brief glitch in understanding, but a second or a longer one will destroy your hopes.

For a talk to the department, assume that the members of your audience are intelligent 16-year-olds. For more or less specialised audiences add or subtract a few years to age in the range 12 to 21. Do not be afraid to repeat the main points in your talk: introduce, summarise (say what you are going to say), explain (say it), recapitulate (say what you have said).

This will take time, so start planning by realising that the amount you can get through in a talk is limited. It is much better to get one idea across firmly than to deliver two ideas neither of which is understood or remembered. How long are you allowed for your talk? The organisers of the event you are going to should have told you: if they have not then ask them. An account of a bit of primary research will usually be allotted 15 to 30 minutes: 20 is probably the commonest. An invited 'seminar' or 'plenary lecture' will be expected to cover more ground and may be allowed 40 to 60 minutes. (Longer times are not an unmixed blessing as there is more opportunity to alienate the members of your audience: some of the ineffective talks I have sat through began well but after a few minutes accelerated into the intellectual stratosphere leaving even the experts behind.) In what follows I concentrate on the talk to a moderately specialist audience.

Expect to spend much more time preparing your talk than the time allowed for it.

It is usually helpful to think first of a single theme or thread or storyline that you will follow. Then you can use that to decide what visual supports you need.

Allow 20% of your allotted time for questions and discussion at the end. For a 20-minute slot you will then have only 16 minutes for presenting. And if there is to be serial translation ▼ you will have only half the remaining time: 8 minutes at most. Deciding what to leave out can be more difficult than deciding what to put in. Limit your scope. Prune without mercy.

> ▼ 'Translate' tends to be used for the written process and 'interpret' for the oral (spoken) one, but the words have overlapping meanings and at international meetings the oral process is often described as translation. It can be of two kinds: 'simultaneous' ('parallel') or 'serial'. Simultaneous translation is a highly skilled and, for the translator, stressful undertaking. It consumes none of your allotted time but, a day or so before your talk, you should provide the translators with a list of the technical terms you are going to use and their meaning. Then 'hydraulic ram' will not get translated as the equivalent of 'water sheep'. You won't usually hear the simultaneous translation. Serial translation is audible to you and the members of your audience: first you speak one or a few sentences, then the translator speaks, then you speak again followed by the translator, and so on alternately. The serial translator may be a professional, but is more often a colleague skilled in English and another language and knowing at least something of the subject.

Most talks about primary research follow a pattern similar to that of the Summary of a written article.

- First, a skeletal introduction (why the work was done) followed by a brief one-sentence summary of what emerged. Be careful to set the scene. Remember the army adage: 'Time spent in reconnaissance is rarely wasted.'

- Next is a brief account of any essential non-standard or unobvious method. (If your hearers want to discover more about the methods they can ask in question time or after the whole session ends.) This may take a minute or two.
- Most of the rest of the time is available for results and their interpretation. This section should differ from a written account in that it is often, perhaps usually, necessary and sensible to combine what would in a written version be separate Results and Discussion sections. If the work was logically sequential then arrange what you say like this: first question A, results and discussion of A leading to a second question B, results and discussion of B, alternating question and answer as many times as necessary.
- Finally, you should present a clear summary containing the 'take-home message'.

If all this is done well the members of your audience will have found your talk interesting, there will be stimulating questions, and the chairperson will have to stop the vigorous discussion (to allow the next speaker or the tea break to start).

(3) Visual aids

What you say provides the necessary links between the evidence, usually presented as a computer-controlled projector 'presentation' (CCP). Older methods – blackboard, passive whiteboard, paper flip-chart, overhead projector, slide projector – are less flexible and have become uncommon. A new apparatus, the computer-linked whiteboard, may be available. It has some of the advantages of CCP and allows input from the speaker or members of the audience. It is useful for small discussion and planning meetings, but less than CCP for ordinary scientific meetings.

Do not forget the blackboard and passive whiteboard completely. They need only chalk or the special marker pen, and do not need electricity. I recall a speaker about to give a run-of-the-mill talk, who discovered that his photographic slides had been removed, unintentionally, by a previous speaker. With this stimulus the bereft speaker extemporised brilliantly, sketching rapidly on the blackboard the essence of what his most important slides would have shown. Three decades later I can still remember the main points of his talk.

Nowadays, CCP is widely available. At its best it produces high-quality images. A few years ago both the projector and the computer program controlling it were unreliable, but both have improved. The CCP projector compensates for tilt so the projected image is rectangular, whatever the angle of elevation of the projector. It is easy to assemble slides – the name 'slide' persists from the early days of slide projector where the images really were slid into place. It is easy to use them flexibly during a talk, and to build up complex diagrams and lists in stages. And you can use short films and other special effects. You can misuse them too.

Most modern laptops support 'hot switching' in which you have everything running and just plug in the cable to the projector and start within a few seconds. This would allow you to use your own presentation program. But the organisers of your meeting may insist on all CCP talks using the Microsoft PowerPoint format ('.ppt' and '.pptx' file suffix) and that your file is loaded on their computer before the session begins. This tether to a monopolistic commercial enterprise is distasteful, but is for practical reasons. The next speaker may need to use the organisers' PowerPoint program and after you have finished it may take one to several minutes to get that program going. There is simply not enough time in a busy meeting to allow you to do this between talks, and changing the image on the screen while the previous speaker is answering questions is distracting and bad manners. The safe choice for a smoothly run session is to have all files pre-loaded on the organisers' computer.

Slide order and design

In a written article the tables and figures are the evidence you write about. But in a talk the visual illustrations are *aids* to the story you are going to tell. The most important slides are the first two and the last.

Start with a slide that lists the title of your talk, your name, your institution, and the names of those you need to thank. This starter slide is sacrificial: it contains nothing that the members of your audience need to know that they do not know already from the meeting programme. Its purpose is to let people settle down and get used to your appearance and voice. This slide is the one place you might justify a decorative background. Don't talk about the slide. Show it for perhaps 10 seconds. Follow with the most important slide of the talk, the one which grabs the attention and interest of the members of the audience. Once they are hooked then the battle is half won.

Finish your talk clearly with the last slide showing your 'take-home message' on it. That slide will then remain there for the members of your audience to absorb during the questions and discussion. A common mistake is to end the talk with a long list of everyone who had anything (or nothing) to do with the work. This is mostly unnecessary and wastes time, especially if you insist on reading the names. Further, you leave the members of your audience during discussion to gaze at this largely irrelevant list. Put the necessary names on the first (sacrificial) slide, not the last.

A background picture will distract the members of your audience, even if it is relevant to what you have to say. Particularly irritating is a logo or other advertisement for your organisation, repeated on every slide. Too many organisations are infected with a publicity department that 'requires' such distractions. Avoid them if you can. An effective talk is the best advertisement your organisation can get.

The CCP allows you to choose 'transitions': the way in which you move from one slide to the next. You can, for example, fade, flow or fly in from top, bottom, or either or both sides, pixellate in a pattern or randomly, or rotate dizzyingly. All these effects are there mainly because they can be programmed not because they are needed or useful. All are distracting, none (in my view) has any merit in a talk about science. If you must use one of these unusual transitions then stick to one only, and use it consistently and considerately.

Detailed suggestions about slide design follow.

(i) The '.pdf' format seems to be the most generally reliable one for inclusion in a '.ppt'.

(ii) Use the whole area of the slide.

(iii) An optical illusion makes a bright screen in a dimly lit room look bigger and brighter than it really is. If you sit 9 m away (at the back in a 60-seater lecture room) from a screen 150 cm square the picture subtends 9.5° equivalent to a 10-cm square of paper held at arm's length. The detail that will be distinguishable by someone at this distance is limited. Figure 2.1 is overcrowded. It takes perhaps 10 seconds to read, during which time, if you continue talking, the members of your audience will not be listening to you. Figure 2.2 is better (but contains a spelling mistake, thus creating a distraction and a bad impression). As a guide for text, use no more than 25 words in Helvetica or Arial font, on 10 or fewer lines. Print the slide to be about 10 cm square: it should be readable at arm's length.

(iv) In text slides, use no more than two fonts or faces on a single slide or in the whole talk. Figure 2.2 shows this (*'Example'* in italic face is the second face). If you use more you divert the attention of the members of your audience who will be wondering about the significance of the different fonts.

(v) Be careful with Latin names of organisms and chemical names, both of which are peculiarly difficult to grasp when heard for the first time: show them spelled out on a slide too at first mention.

(vi) Foreground black contrasting with background white is the usual choice for text slides, but the contrast of black text on white on a large screen in a dimly lit room can be uncomfortable: yellow against dark blue gives a less violent contrast between foreground and background, and between room and slide. The human eye is most sensitive to yellow so yellow on blue is better than blue on yellow: light text on a dark background has a larger signal : noise ratio, and some people have difficulty focusing blue letters. Bright yellow on medium green can be a good combination too.

(vii) Use colour judiciously, not just because it is possible, and use it to distinguish different types of material, not arbitrarily for

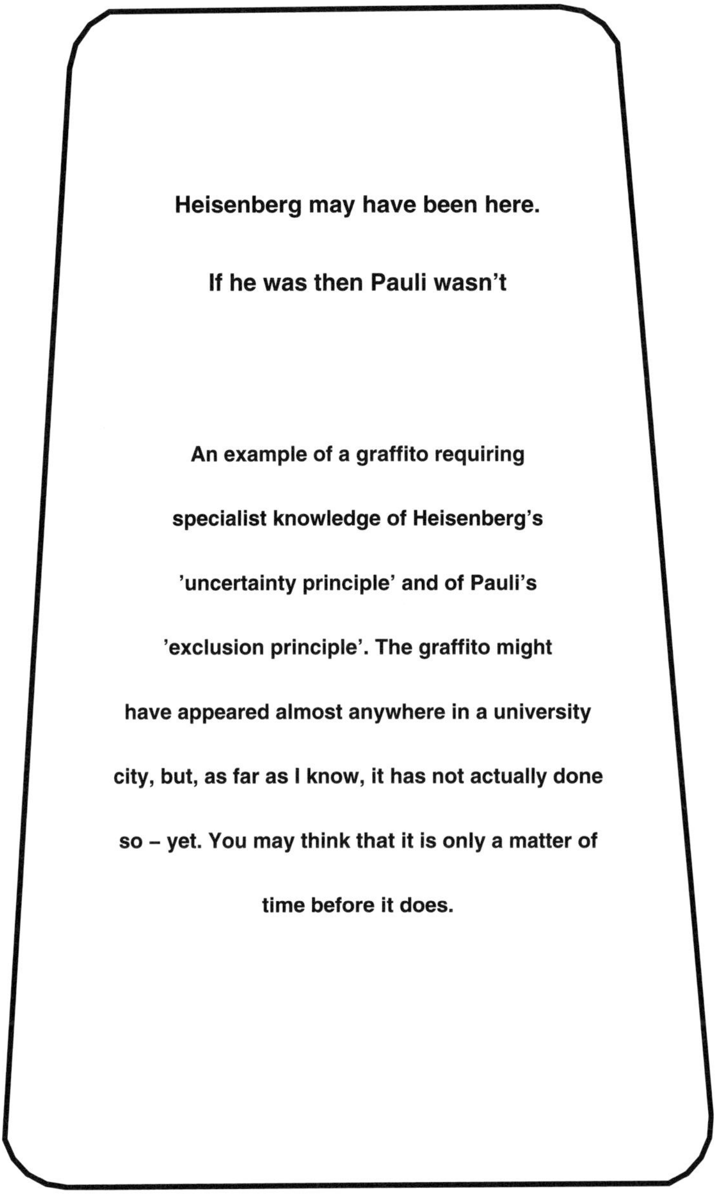

(3) VISUAL AIDS

> **Heisenberg may have been here.**
>
> **If he was then Pauli wasn't**
>
> **An example of a graffito requiring specialist knowledge of Heisenberg's 'uncertainty principle' and of Pauli's 'exclusion principle'. The graffito might have appeared almost anywhere in a university city, but, as far as I know, it has not actually done so – yet. You may think that it is only a matter of time before it does.**

FIGURE 2.1 A badly designed, near useless, slide. Too much text and font too small. Boundary too obtrusive. Why the unusual shape of the boundary? Compare with Figure 2.2.

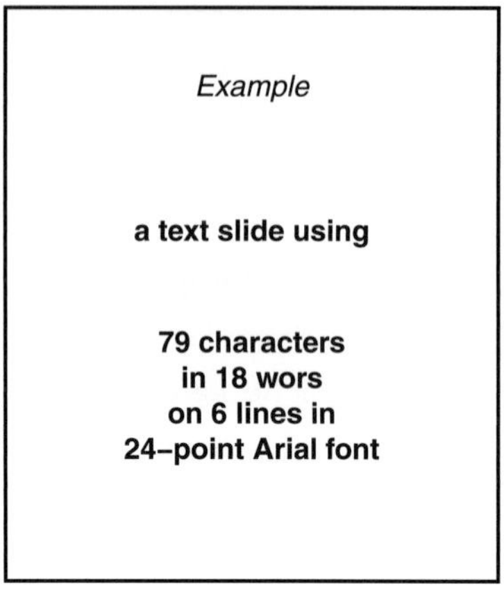

> *Example*
>
> **a text slide using**
>
> **79 characters**
> **in 18 wors**
> **on 6 lines in**
> **24–point Arial font**

FIGURE 2.2 A simple slide with easily readable text. Notice how the error in 'words' distracts your attention.

artistic effect. Apart from the main foreground colour and the background against which it contrasts use no more than two other colours, and preferably none. More will divert the attention of at least some members of your audience who will be trying to understand the logic of your choices meanwhile missing what you are saying.

(viii) The colour of thin lines cannot be distinguished by people at the back of the room.

(ix) Remember, too, that colours with the same 'luminosity' (brightness, density) lack contrast so people at the back of the room will have difficulty distinguishing them.

(x) Be aware of red/green colour-blindness which afflicts about 8% of people (mostly men): if you write in red on a green background such people will see only a uniform blank space. Some people find that red itself appears to stick out from the screen. The website <www.vischeck.com> (6 Sept. 2013) allows you to see your slides as someone with defective colour vision would see them

(xi) Some colours seem to enhance adjacent ones: experiment. Avoid orange and blue (some people see them as vibrating against one another).

(xii) If you need to have a break between sections it can be more restful to have a black or dark grey slide rather than a bright white one.

(xiii) Use tables only rarely.

(3) VISUAL AIDS

(xiv) Simplify tables – if you must use them – and figures to the essentials: those designed to be printed will probably need redesigning. A complex interrelations ('box and arrow') diagram may work well in print, as readers can concentrate on a bit at a time. But the same diagram will be unsatisfactory for a talk, because viewers will not have the time to understand it. In figures, you probably need to thicken lines, enlarge text, and (if possible) orient graph ordinate labels horizontally so the members of your audience can read them easily without twisting their heads parallel to the floor. Bar charts need care because the members of your audience may need more time to understand them than they would a simple trend graph. The limits are about 3×3 (three sets of three bars each), 4×2 or 2×4, or 6×1.

(xv) Give each slide a short title describing the contents. Not general such as 'Results' but specific such as 'Larger bees forage further'. These titles in order should approximate a summary of the story you are telling.

(xvi) If you really must use a complicated slide, build it up bit by bit, by adding to an initial simple slide. That is easily done in CCP by duplicating the first slide (perhaps converting existing points in a list to grey) then adding the extra material.

(xvii) A short 'video' can be included in a PowerPoint file and can occasionally be really helpful. But use this feature only when necessary. Apart from anything else such 'video clips' are still technically unreliable. They worked when you rehearsed at home, but will they behave at the event itself? The sight and sound of your struggle to get a clip running takes time, is unedifying, and destroys the flow of your argument. Simpler 'animations' in which a small part of the image makes a repetitive cycle of movements are just as distracting. The human brain is adept at noticing such small changes, perhaps because this ability has strong survival value.

When you have designed your slides, assemble the images you will need to import to the presentation. The Appendix to Chapter 10 describes the features of RGB colour mode (that you will need) and the resolution needed. For a talk, anything over 50 dpi (dots per inch) should be plenty. While the publication may need images in large '.tif' files the same image can be put in a much smaller '.gif' or '.png' or '.jpg' file for the talk with a great saving in file size and hence in speed of presentation.

Finally, three general points.

1. Will you need 'speaker's notes'? If you write out at length what you want to say you may be tempted to read from your notes. Resist this: you will probably lose the members of your audience. Your visual material will be sufficient to prompt you. If you are still doubtful then write brief 'prompt' notes on cards, punch a hole in the corners, and tie them loosely with string (to avoid losing the order if you drop them). Then put this comfort pack in a pocket and forget it.
2. It has become common practice to carry the PowerPoint file on a USB 'memory stick' (courageous speakers may even download their file over the internet). These 'sticks' are now cheap, so make a copy of the file on two separate 'sticks' of different manufacture and perhaps on a CD or DVD disk too, against the event that your primary 'stick' has been corrupted. And check that you are prepared to extemporise on a blackboard or whiteboard if the projection system fails.
3. It is salutary to consider beforehand how you could transmit the essentials of your talk in half the time, because an incompetent chairperson has allowed speakers to overrun their allotted time leaving you to compress your talk. I have several times had to cut a talk designed for 20 minutes to less than 10 'on the fly'. On every occasion the talk has been unusually effective, because there was no time for the usual protective qualifications. Why have I never learned this lesson?

(4) Practising and refining

Before you go to the meeting, perform your talk at least once or twice to a small audience of critical friends and colleagues. (Have a look at the later section 'The performance itself'.) If you can arrange to be videoed do so: the results will almost surely be illuminating. Perform the talk in full and note how long it took. You may then have to remove something to fit the allowed time.

Your friendly audience will tell you if you are inaudible, or are gabbling, or cannot be understood, or ... It is much better to get such criticism before the real performance than after it.

During practice you may realise that particular slides add too little to justify the time they consume. You may also get ideas for how to improve the story.

Giving a talk is the most exposed of the usual ways of communicating science: you are on your own in real time. Confidence comes with practice and experience. Some speakers, particularly the inexperienced, prefer to rehearse repeatedly, to an empty room when necessary, as actors do to deliver a set script. Some more experienced speakers think that perhaps two rehearsals are sufficient. They are then extemporising words and sentences from knowledge of the whole structure of their talk, rather than from a set text, and can therefore adapt more effectively to unexpected circumstances. A few experienced speakers do not rehearse at all, and the quality of their performance too often reveals this.

All preparation being complete your attention turns to the event itself. What will you wear for your talk? In general, dress a bit more formally than you think the audience will have done. For a talk at a learned society meeting dress soberly, to avoid distracting the members of your audience. But excessive formality, especially in grey, tends to induce sleep, and a talk within your own institution will usually need less formal dress. A dash of bright colour relieves monotony in all circumstances. Lapel microphones, if you have to use one, may not fix well onto flimsy fabrics, and the transmitter of wireless microphones may need a proper pocket – both can cause difficulties for women.

(5) On the day but before your talk

Most meetings nowadays are built around CCP and specifically PowerPoint files. Meeting organisers may require you to copy your file to a central server that is then used by projectors in all the lecture rooms. Others expect you to copy your file to the organisers' laptop specific to, and physically in, the room you are to speak in. Whatever the system, make sure you know what you have to do and if it needs action before the session begins do what is needed in good time for your session of talks.

The chairperson should make sure before your session of talks begins that the physical conditions are as good as can be got in the circumstances. But for your own good arrive 15 minutes before the start of the session in which you are to perform, and do this yourself.

- Is the room heating suitable and under control?
- How do the lights and window blinds work? In particular, how do you set the lights to 'dim' and 'full'.
- How do you control the projector?
- Where is the pointer you intend to use?
- Clean the blackboard or whiteboard if it needs it and remove any distracting loose papers, empty water bottles, and other litter from the front of the room and the lecture bench.
- Introduce yourself to the chairperson (who should be anxious to know that all the speakers have turned up).
- Check your first few slides and make sure that you can move easily backwards and forwards directly to a different slide (omitting all the ones between it and where you are). But don't leave one of your slides showing.
- Occupy a seat near the front, take half a dozen deep breaths, and concentrate on what you have to do.

(6) The performance itself

Choose someone in the back row and speak mainly to them (but do scan the rest of the audience too, and speak also to some of those who look particularly interested). Don't speak to the screen or to the computer: neither will take any notice.

Gabbling, and muttering quietly to the computer alone, are the almost universal responses to stage fright. Here are the three cardinal rules:

| *Face out.* | *Speak up !* | *Slow down ! !* |

(6) THE PERFORMANCE ITSELF

Skip your sacrificial first slide after a few seconds when silence has settled. Then show and talk about your crucial second slide – the one that grabs attention.

A first repeat: speak slowly – much more slowly than seems natural. (If English is, for you, a learned rather than a native language then be specially careful to speak slowly.) Don't gabble. Leave pauses. The pauses while you assemble your next thought can seem to you to be embarrassingly long. But in reality they are probably short and welcomed by the members of your audience as it gives a second or two for what you have just said to sink in. If you have videoed yourself in a practice performance you will already be convinced of this.

Let your voice rise and fall. Vary your pace, slow over a difficult section, faster over easier ones. Use emphasis occasionally.

Use simple grammar and words – the members of your audience cannot look things up in a dictionary – and avoid jargon that the members of your audience are not familiar with. Jargon is fine when used between informed communicants, but not if it obstructs understanding by the inexpert.

Enthusiasm is infectious and will help your talk along.

Some people think science is a serious business (it is) and that talks about it should be as well (not so). A leavening of humour, especially after a difficult section, can have a good effect allowing the members of your audience to relax for a moment before the next steep ascent. Humour is always risky, and much of it does not travel across national boundaries. To be effective it should be relevant, brief, 'suitable', and trans-national. If in doubt, leave it out. Seek advice after your first or second rehearsal.

Do not read from slides: the members of your audience can read more quickly to themselves than you can read aloud to them. You are supposed to be leading them, not following. But it can be useful to show a list of trigger words and then amplify or comment on them orally.

Always explain graph axes. You should already have ensured that labels will be just readable at the back of the room. (Could you read them when the slide was printed on a 10-cm-square sheet of paper and held at arm's length?)

It is a surprise that colour seen by the members of your audience, some way away from the screen, is often different from that seen by you, the speaker, close to the screen. For example you see red while the members of your audience see brown, green to you looks yellow to them. On average up to 8% of the audience will be red/green colour-blind (you should have avoided this problem during preparation). Never refer to a feature on the slide by its colour: point at it or describe it ('the top right point') instead.

Don't flash a slide on and then remove it before the members of your audience can grasp what it is about. The temptation to do this will be less if you always speak *and* point to the important features on slides.

Take special care to avoid teleology (imputing design to the work of evolution as in 'The male animals have longer necks to see further').

Avoid 'tics' of speech (repeated 'er', 'um', 'like', 'I mean', 'sort of', 'OK') and avoid dropping your voice to a whisper at the end of each sentence. Avoid physical 'tics' too, such as pacing about and about, tapping your fingers on the bench, waving your arms about (except occasionally to emphasise a point). If you are a natural 'pacer' try putting down a sheet of newspaper and giving your whole talk while standing on it. These imperfections should have been removed by sufficient rehearsal.

Use a microphone? Preferably not. If the room seats 150 or fewer people and you speak to the back row you should not need a microphone. If you do use one then speak in your normal voice, not raised, and try to stay the same distance from the microphone. If it is fixed to your person this is easy (but mind the cable if there is one). If the microphone is fixed to a stand or lectern then you must stay next to it – a good reason for not using it. If a loudspeaker begins to howl (because the microphone is picking up sound from the loudspeaker itself and then re-amplifying it) shield the microphone from the loudspeaker with your hand, or

(6) THE PERFORMANCE ITSELF

abandon the microphone altogether. There may be a technician who cures the problem by turning down the amplifier volume control.

A simple mechanical pointer, if practicable, is probably the best way of locating something on the screen, but organisers may discourage mechanical pointers lest they damage the screen. You do have arms and fingers, which may be enough on a small screen. Use a laser pointer? Most are too faint, and even small nervous movements of your hand become wild wanderings on the screen making the spot on the screen almost impossible to see. The mouse-controlled pointer on the computer screen may be better because it doesn't shake so much. For a large screen, say 5 m across hung with its base 2 m off the ground, the mouse pointer may be the only effective mechanism. You can always help by describing where you are looking: '... the graph in the top right corner ...'. If you do use a laser pointer, avoid pointing it at the audience.

Introduce your last, 'take-home message', slide with 'Finally ...' to show you are at the last slide. (Use 'Finally' only once!)

If you have finished when you should have done there will be time for a few questions and perhaps discussion. Do not fear questions. You know more about the subject of your talk than anyone in the audience does. It is generally agreed etiquette that one does not seek to embarrass a speaker giving their first few talks. If there is serious controversy it should be among more experienced members of the audience. Answer questions succinctly and stick to the point. If the questioner seems to be the only one who has this bee buzzing in his or her bonnet then be brief. The aim is to be helpful but to allow as many of the members of your audience to take part as possible.

Three reminders: (1) Don't leave your audience in total darkness. (2) Keep to time! Leave 20% of your allotment for questions. (3) For the last time:

Face out.   *Speak up !*   *Slow down ! !*

Finally: good luck.

> **How to lull your audience to sleep, adapted from Booth (1993).**
> - Wear a dark suit and an inconspicuous tie.
> - Make sure the lecture room is warm and dark (lights off, close the windows and blinds).
> - Arrange to begin your talk about half an hour after lunch.
> - Show that you are not interested in your subject.
> - Show crowded slides without explanation, especially of graph axes
> - Stand still and read your talk in a fast whispered monotone, without looking up, and without emphases or change in speed
> - Show few slides and remain completely serious throughout.
> - Use grandiloquent words and unsuitable jargon in long, complex sentences.
> - zzzzZZZZZZZZ zzzzzzzZZZZZZ zzzzzzZZZZ

QUESTIONS:
- Were there questions after your most recent talk?
- If so, what did you learn?
- If there were no questions, then why not?

3

Making and displaying a scientific poster

As scientific meetings got bigger some attendees were not able to give a talk because there was not enough time. The scientific poster was invented to allow these disadvantaged persons to display their work.

Poster sessions have been common for only a few decades, so procedures and poster design techniques are still evolving. The poster is the least formal of the three main communication methods, and corralling viewers for it is competitive. It has the same relationship to the formal article as a painted miniature does to a Vermeer portrait. The advantage of presenting a poster (at anything other than a small meeting for specialists) is that you have an opportunity to interest people outside your special field. It will be clear then that visual presentation is the key to designing a poster that gets noticed. Of the three forms of presentation the poster, as the name implies, is the one that has the largest element of deliberate advertising in it. You need to attract attention. Of course the substance of what you present must be interesting too.

The aims of a poster are

to attract viewers,

and the now familiar one:

to communicate with those viewers as clearly and simply as possible.

There are usually five phases to presenting a poster: composing a header and writing a summary; preparing what illustrations and text to put in the poster; designing the layout and presentation details; preparing to leave for the meeting; and setting up at the meeting and performing at the dedicated poster session (there usually is one).

(1) The title and summary

As with a talk and long before the meeting, the organisers will probably ask for a title and summary to be printed in the meeting programme. As for a talk, the length of summary may be set in the range from 50 words to one or two A4 pages. This is the time to give serious thought to what you want to present. The poster is not a formal record: its purpose is communication, to interest a viewer in the work and direct them to where they can find detailed information. As with a talk, resist the temptation to plan in anticipation of results that you do not yet have. The questions are the familiar ones. Who are the viewers you want to reach? What do you want them to remember? The amount you can hope to communicate is even less than you can in a talk: if you bite off too much your viewers will just walk by. Limit your ambition. It is better to put across one simple idea than to fail with something more complex. Some topics are simply unsuitable for a poster.

Even if the organisers want only a short summary it is worth preparing a one-page A4 summary too, with text and one or two graphic images, to be used as a handout at the meeting.

The contents of the summary should, ideally, follow the now familiar lines: concentrate on the question you wanted to answer, what you found, and its implications. Be as numerically specific as you can. But, as with talks, if you are going to show current work then be as unspecifically general as you decently can.

Think about the title. It is the sharp point of your attempt to attract attention. For a formal article explaining why peat accumulates unexpectedly fast at lower temperatures I used the title 'Effects of temperature on the rate of peat accumulation'. But for the poster this became 'PEAT IN A COLD CLIMATE', which is short, mildly intriguing ('What can that be about?'), and parodies the well-known Nancy Mitford novel *Love in a Cold Climate*. (This title itself came from George Orwell's novel *Keep the Aspidistra Flying*, which might, in suitable circumstances, be parodied for a poster title. The forces that produce memorable novel titles are the same as those you face for your poster title: competition and advertising.) Of course the title must be reasonably related to the poster subject. Avoid jargon and two-part titles with a colon. Restrict the title to what will fit on a single short line on the poster.

The main point is that informality is widely accepted in posters, and a declarative title summarising the message will not cause the aversion that it might in a formal article: you could get away with the poster title 'Malaria is doomed' or even 'The Moon is a blue cheese'.

(2) Preparing the poster content

The poster relies primarily on graphic images (such as photos, graphs, diagrams) linked by the minimum of text. *The format of an article intended for publication, or already published, is totally unsuitable for a poster.*

Assemble the graphic components first. The appendix to Chapter 10 gives advice about colour mode, resolution, and file formats. If you can, get advice from a professional designer. Here may be a use for someone in your publicity department. As the poster is to depend on external illumination then CMYK would be best, but in practice RGB is usually easier and allows one to use file formats such as '.gif' and '.png' and '.jpg'. Graphic objects such as '.jpg' images that have pixels (not vectors) should have a resolution on the printed poster of about 300 dpi (dots per inch). Drawing programs allow you to print your design at a different size than it

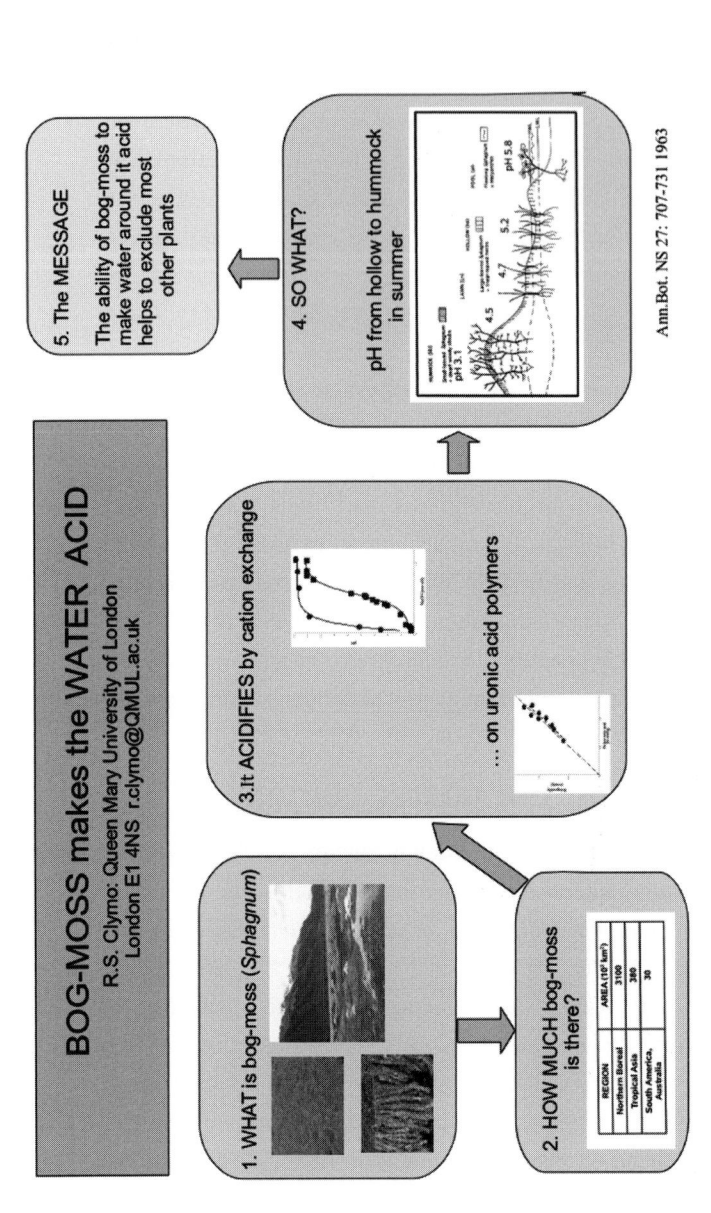

FIGURE 3.1 Intermediate stage in the design of a poster. The explanatory 18 point size text in the panels has been omitted because it is too small to read at the reduction necessary for this book page size. The original colours appear here as shades of grey. Notice: landscape orientation; strictly limited scope of the subject; informal language and sections; short title; uppercase for emphasis; title and headings sanserif font; only three type sizes (two visible, the third for the omitted text); navigation from left to right by numbers and arrows; summary ('The MESSAGE') is at the top right, level with the title and visible over visitors' heads; panels of different sizes; rounded corners; varied graphic objects (three photographs, one small table, two simplified graphs, one line diagram); title panel contains author, institution and addresses; no background pattern or logos. The poster is designed for a single A1 sheet, but could easily be produced as six small sheets, fixed separately to the display board with linking arrows also fixed. WARNING: this poster is just an example. It is NOT a template to be copied uncritically.

was when you designed it. So it is common to design at A4 size but to print at A1 size, three sizes larger, so everything is enlarged by a linear factor of $(\sqrt{2})^3 = 2.8$. So an image on the A4 size must be at $2.8 \times 300 = 840$ dpi. (The 'A' sizing system is explained in the next section.) Internet images often have resolution of only 72 dpi and may be near to unsuitable for a poster.

Graphs produced by default settings in a graph-plotting program are often unsuitable (Chapter 10). Suppress shadows, excess grid lines, and other fancy decorations. Apply the same or similar text size rules to axis and other labels as you do to text (see later). Simplify mercilessly.

If you cannot avoid tables then simplify them unmercifully.

Then compose a minimal text to link the graphic materials. More than about 350 words is too many. This is a common fault and is seriously repellent. In attractive posters graphics often occupy three times as much space as text. The text of an effective poster usually has a clear and simple story line. Use active constructions not passive: 'This shows that ...' rather than 'It was shown by this that ...'. Use short words and be concise.

The essentials are to explain first, what the purpose of your poster is, the question you are answering, why it matters, and (perhaps, if not obvious) how you set about it. Recall that you can hope to communicate less even than in a talk. The purpose can even be obvious in the title, as it is in Figure 3.1. The headers in a poster may be less formal than those in an article, even though the poster content may cover similar ground. If you start at the top left of the poster you don't even need to label this section 'Introduction', or you might label it 'Purpose' or 'The problem'.

In a conventional article the methods are collected together (and often appear at this point), and they are mentioned briefly even if they are standard ones. In a poster, however, you may be able to do without any mention of the methods if they are standard or obvious ones. If someone is interested in details they will ask. If the results will not be understandable without a lengthy description of methods then you may have chosen an unsuitable subject for your poster.

The largest part of the poster should (usually) be the results presented in graphic or very simple tabular form, and each (with its caption) as far as possible self-explanatory. Again, unlike an article, you can comment on the significance of a result immediately after showing it, mixing results and discussion in a way that would be unsatisfactory in an article. If you do this you may not even need a separate discussion at the end.

You may be tempted to have a section headed 'Conclusions' at the end. If the logic of what you have put in the poster is so complex as to justify that then you have probably chosen an unsuitable subject for your poster. But the urge to write such a section may be simply that you recognise the need to write a 'take-home message'. In a poster this takes the place of the Summary and probably of the wind-up at the end of the Discussion in an article. Where to place it physically on the poster is a matter for the design and layout.

(3) Designing and laying out the poster

Here we are at the heart of creating an effective poster. Some of the topics that follow command respect by force of reason, others are open for discussion and personal judgement. Remember that in the formal poster session there will probably be a crowd, its members (possibly with a glass in hand) drifting slowly by. Your poster has to grab their attention in the few seconds it is within sight, so it must be striking, easily readable from an arms-span away, and must *appear* to be easily assimilable in only a few minutes. One strategy is to produce a very simple poster with only one or two striking large images and very little text, but to have the real poster on A4 handouts. Study the professional salesman's techniques.

(i) A common size of poster is on a single large A1 sheet (84 cm × 59 cm) ▼. If in the 'portrait', ▯, position (with the long axis vertical) then the 84 cm distance from top to bottom is uncomfortably large for most people: either they have to crick their necks to read the top (particularly if they have bifocal spectacles) or stoop to read the bottom. The maximum comfortable vertical dimension is about 65 cm. So, design your poster for the 'landscape', ▭, position. Aim to set the centre of the poster substance (excluding the title bar) at an average eye-height of about 150 cm.

> ▾ Why the peculiar size? The A0 (119 cm × 84 cm) area is 1 m^2. The length : breadth (aspect) ratio is $\sqrt{2} : 1$. The value of this ratio was chosen so that when you cut the page across in half you produce the next of the A series (A0, A1, A2, A3, A4 and so on) each with the same aspect ratio as the original. The long side of the next smaller size is then the same as the short side of the size it derived from. The quotient of the short / long sides is 0.707, not far from the 'golden ratio' 0.618, which is widely held to be aesthetically pleasing.
>
> The B series has the same aspect ratio as the A series with dimensions that are the geometric mean of the A series: B1 dimensions are $\sqrt{(A0 \times A1)}$. The C series is similarly derived as $\sqrt{(A1 \times B1)}$, and is used mainly for envelopes for the other two series.
>
> Single-sheet posters are moderately expensive, and A1 is probably large enough for most purposes though many posters are A0 size. I assume A1 size for simplicity. Both A1 and A0 need the same size of type. The chief advantage of A0 is its larger area allowing more material to be displayed, but this can easily look forbidding to a viewer; better to increase the proportion of space.

What if the organisers supply only narrow display boards, less than 84 cm wide? One solution is to hope that it will be practicable to have your landscape poster projecting a bit from the edge of the display board, possibly with light supporting stiffeners. Another solution is to print at A2 size, though this does severely restrict how much you can fit into the poster. A third solution is to design to the maximum width allowed by the display board with a corresponding reduction in height to no more than about 65 cm.

A fourth solution is to design the poster in small (say A4) chunks that can be arranged to suit the display board dimensions. The last, and worst, solution is to design your poster for the A1 size in portrait orientation, as the organisers expect you to do.

(ii) The single-sheet poster is convenient, even if a bit awkward to transport. Since fear of terrorist attacks made searches at airports more strict you may find that you are not allowed to take a rolled poster on to the plane yourself but have to consign it as luggage (where it is easily damaged or lost altogether). Some airlines will charge exorbitant amounts for a poster tube as 'hold' luggage, and even worse for cabin luggage. Posters as small chunks are easily carried in a bag. Or you may be able (a newish technique not available everywhere yet) to get your poster printed on flexible material that can be folded for transport. Creases can be smoothed out when you arrive by ironing the reverse side of the fabric.

(iii) Consider the whole poster as the display 'arena'. An undivided single-sheet poster is fairly forbidding, so group items in half a dozen discrete blocks or panels. These panels should be clearly delimited either by a border or, usually better, by a background colour different from the arena colour. The panel background should be a light colour (or white) because you will be filling the panel with graphics and text. Avoid a dark arena background: it makes for distracting contrasts. White and light yellow are often used for arena and panel (or reversed). Light grey emphasises colour in photos while white diminishes it. Some people think that rounded corners to the panels are aesthetically more pleasing than square ones (Figures 3.1 and P.1 in Preface). The aim is to delimit the panel clearly but not so vigorously as to be distracting.

(iv) The panels can be, and probably are best to be, of differing sizes, but (with the exception of the title and perhaps the take-home message panel) should be of consistent design. This helps viewers to understand easily where they are in your story. Most design software allows you to set a design grid, and this will ensure alignments among panels, where these seem useful.

(v) People scanning the poster need to be shown the order in which to scan the panels. In general a viewer should need to move only once across the arena from left to right. If the panel order runs from left to right along a top row and then back to the left for a second lower row the viewer has to cross with someone else who is still on the top row. Indicate the order with sequential numbers in the panel header, or with conspicuous arrows, or both (Figure 3.1).

(vi) Use colour sparingly (apart from photos and graphs) and consistently, and for a specific (preferably self-evident) purpose. Arbitrary splashes of different colours for no obvious reason simply confuse your viewers. Bright primary colours are the visual equivalent of shouting. Arena background and panel background, and perhaps a third, bright, colour for headings, are really all that should be necessary. Use black for text. Recall that on average 8% of people are red/green colour-blind and avoid this contrast (which is poor for the other 92% too) as are brown with green, and any pair of blue, black, and purple. In general avoid adjacent colours of the same luminosity (roughly, those which have the same greyscale when printed on a black-and-white laser printer).

(vii) The text must be easily readable an arm's-span away. That implies text characters that are at least 18 pt, and preferably larger. Panel headers should be 40 pt, and the title bar 100 pt. Serif fonts (such as Times New Roman and Liberation Serif) were designed for lines of text and should be used for them. The default on some drawing programs is sanserif. Change to serif for text. Sanserif fonts (such as Helvetica and Arial) were designed for headers and short phrases: use them for these only. It is worth printing out the text from a panel separately and seeing if you can read it from an arm's-span away. If you, knowing what is on the panel, cannot read it easily then you need a bigger or different text font. Choose one sanserif and one serif and use only these. Avoid all unusual and fancy fonts.

(viii) Experiment with the line spacing of running text. Slightly more than single line spacing may make the words more readable. Left-justified text is easier to read than fully justified because the space between words is always the same. Use consistent styles for layout: for example headers in the same font and size and with the same space above and below their text.

TABLE 3.1 Adjustments to suggested font sizes and resolution of pixel-based graphics if designing at smaller sizes for printed display at A1 and A0 arena sizes. For example, designing text at A4 size for A1 display then use 36 pt for the title; designing at A4 for text on a part of the poster printed at A3 size for an A1 arena then use 13 pt text; designing a '.jpg' image at A4 for A1 display then use resolution at least 850 dpi

A series of linear steps to display size	A3	A2	A1	A0
An / A0 factor	2.828	2.0	1.414	1.0
E.g.	A4 to A1 arena A3 to A0 arena	A3 to A1 arena	A4 to A3 parts for A1 arena	A4 to A4 parts for A1 arena
Component	Font size (pt)			
Text	7	9	13	18
Heading	14	20	28	40
Title	36	50	72	100
	Pixels (dpi)			
	850	600	420	300

(ix) If you are designing your poster at a smaller size than it is to be printed at, then recall that each step in the A series of sizes increases linear dimensions by $\sqrt{2} = 1.414$. In Table 3.1 are adjustments to font sizes to achieve a satisfactory size on the printed poster (or its separate parts if not unitary).

(x) Put the title in its own panel, distinct from the poster content ones, at the top left of the arena. Include with it the author(s), affiliation plus postal address (combined if possible), and email address. This panel should be visible over the heads of the members of a crowd, and this is the one place where it can help to have a bright primary-coloured background.

(xi) Consider the use of capitals and lowercase in title and headers. In text and graph labels one expects an initial capital and then all lowercase (except where convention requires capitals, for example 'DNA'). The point about the capital is that it attracts attention to itself. Used to indicate the start of a phrase or sentence a capital is helpful, but used to start every word or every 'important' word in a phrase is just distracting. A title all in capitals is more difficult to read than one with an initial capital and the rest lowercase, but there may be a case for putting most or all of a short title entirely in capitals for emphasis (Figure 3.1).

(xii) The last, and arguably the most important, element of the poster is the take-home message. This substitutes for a formal Summary and Conclusions. Logically therefore you want it to be both first and last. This apparent impossibility can be achieved if you put the take-home message panel at the top right corner at the same level as the title bar (Figure 3.1). Viewers see and can read it as they approach, and the ordering device can direct them back to it at the end. This also avoids putting the take-home message in the most difficult place to read at the bottom right corner.

(xiii) It can be useful to add a small photo of yourself on the top part of the poster, so that when you are not attending the poster an enquirer will know who to look for.

(xiv) Acknowledgements (if necessary) and references supplement the main substance of the poster. They need not be in a conspicuous panel and can be tucked away at the bottom right corner of the arena as plain text on the arena background (Figure 3.1).

(xv) Avoid irrelevant and distracting decorative images in or on the background and especially on the panels. Avoid logos too: you are trying to communicate not to create a happening. Your institution is best served by high-quality, interesting, work. If you feel obliged to include a logo then the title bar is a useful place, but avoid multiple logos: they increase the distraction and probably irritate viewers.

(xvi) An attractive poster often has proportions of graphics : space : text around 3 : 2 : 1. Does your poster have those proportions?

Design has a large element of personal preference. All the points above are mere suggestions.

(4) Before leaving for the meeting

If you expect to use your poster more than once, consider having it sealed in plastic (laminated) when it is printed. This is moderately expensive and if you are unlucky the surface may reflect glare from the room lights. Laminate only if you need to.

The meeting organisers will (or should) have told you the dimensions of the display board, but they probably will not have mentioned the height of the top of the board. Will it be sufficient to allow the centre of your poster content to be at average eye-height (about 150 cm)? If you have prepared the poster in separate pieces then the layout can be adjusted and this may not matter. But if you are taking a unitary A0 or A1 poster you may be forcing your viewers to stoop uncomfortably. That everyone else is doing so too is not a solution. But you can arrange to take with you some light vertical stiffeners for your poster that will allow you to attach it to the board with the top projecting above the top of the board.

Did the organisers tell you what the board surface would be? Will drawing pins be allowed? You can attach the hooked part of Velcro tabs to the back of the poster (or parts) but will the board have a suitable fabric surface to fix them to? Perhaps a cloth sheet with a top bar and S-shaped hooks at the top that can be put over the top edge of the board would be a sensible precaution? Then you can attach your poster to the cloth. Or perhaps Blu-Tack would work? Or perhaps you will have to suspend the poster with string and S-shaped picture hooks hung on the top edge of the board? The point is that it is worth anticipating problems that with luck will not arise. It is easy to take with you drawing pins, Velcro, Blu-Tack, string, light plastic stiffeners (for upward extensions) and three or four S-shaped picture hooks.

Finally, take with you a supply of one-page handouts that summarise the poster and contain your contact details, and take also A5 sheets of paper, pencils, and a container to hold invited comments and questions left by viewers at times when you could not be at the poster.

(5) Setting up and tending the poster

Mount your poster as early as possible, keep it there as long as possible, and make sure you are there during any timetabled poster sessions. As soon as your poster is displayed, put out some of the stock of handouts so that viewers can help themselves, and some

A5 sheets for comments. Return at intervals during the meeting to replenish the stock of handouts, and to collect comments.

During the poster session stand near but not in front of your poster. Look enthusiastic and welcoming. Some experienced presenters prepare a short (20-second) talk to supplement the poster when answering specific questions from those who are interested. Judging when to speak to someone who seems interested but has not approached you is a skill you develop only by experience. Do not spend a long time talking to one person: keep yourself available.

Some exhibitors have a gimmick to attract attention. If it moves and is relevant it can be successful. Use your judgement.

Lastly, do make time during the meeting (not during the poster session) to tour the other posters with a critical eye on design rather than content. You can learn much from seeing things done well, and even more from seeing them done badly.

QUESTIONS:
- Did you have a good crowd around your last poster?
- If so, why?
- If not, what might you do to improve things with your next poster?

4

Scientific authorship

How many authors does it take to write an article?

The concept of 'literary author' emerged gradually during the eighteenth century, when only a small part of the population was literate, and a print run of 1000 was 'large'. The concept solidified in the nineteenth century with the introduction of powered printing presses and the spread of the ability to read (Johns 2003). Scientific authorship is different from literary authorship (Biagioli 2003). This chapter is about twenty-first-century writers of articles that describe new scientific work or that review already published scientific work. For brevity I call them author, without qualification. There are no simple rules about authorship, and this chapter is no more than a basis for thought and discussion.

What is an author? The answer is not always obvious. Consider the following patterns.

The science of ecology has passed from simple description to complex experiments, analyses, and modelling in less than a century. The British Ecological Society started the *Journal of Ecology* in 1913. As Figure 4.1 shows, the journal has grown about fourfold in 90 years, with hiccoughs during the two World Wars. The Society has also initiated five other journals. The growth in the

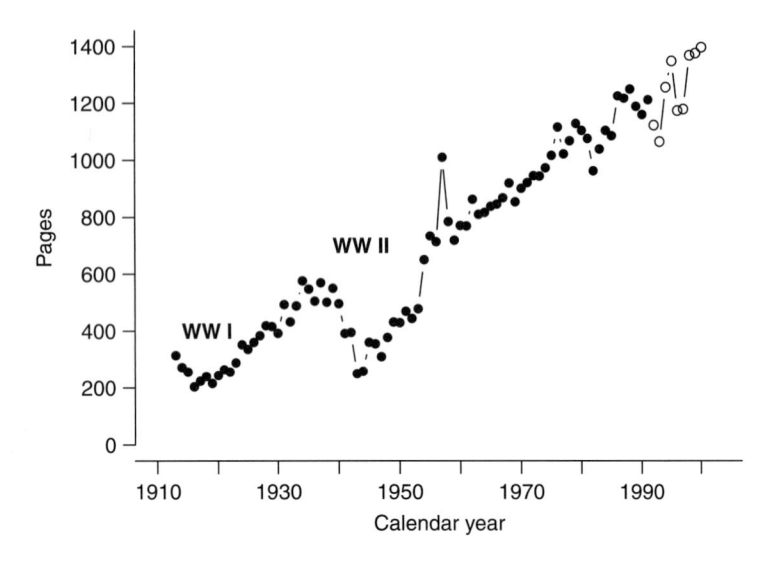

FIGURE 4.1 Number of pages in the *Journal of Ecology* in each year during nine decades. The two World Wars are indicated as WWI and WWII. After 1991 a new format fitted about 25% more material on a page. After 1991, number of pages has been multiplied by 1.25 and symbolised by an unfilled circle.

Journal of Ecology is nowadays controlled mainly by finance. (The graphs stop at 2000 because subsequent production changes make comparisons with earlier years too equivocal.)

Figure 4.2 shows that the mean number of articles in each volume has grown at about the same rate as has pages in a volume, so the mean length of an article has remained about the same. Up to the late 1940s, articles were mainly descriptive and their length increased. After that experimental work and analyses dominated, and the average length of an article stabilised at the smaller value of about 18 pages (longer than in many other sciences).

That ignores changes in style: a modern article contains much more, and more detailed, information than does one published in the 1920s. The biggest change has been in the number of authors needed to write an article. Figure 4.3 shows the decline of the solo effort and the modern rise of group work, implying distributed responsibility, at least partly because ecology (like most sciences) has become more specialised.

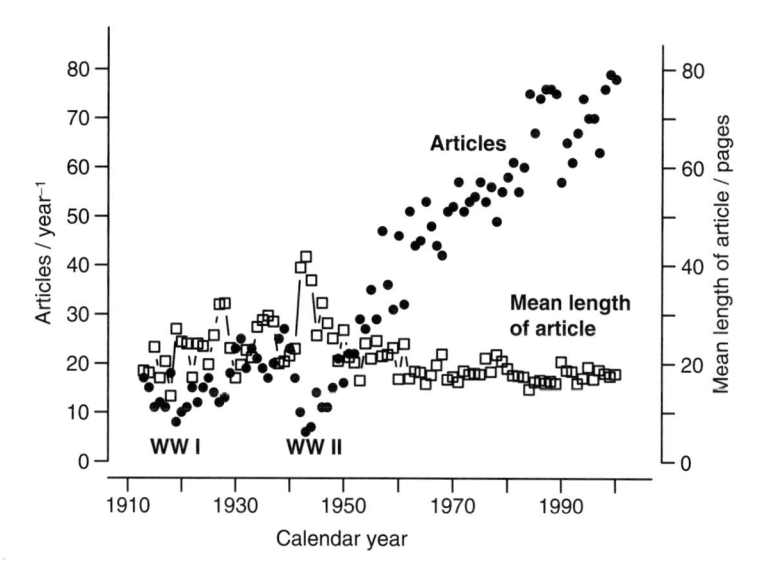

HOW MANY AUTHORS DOES IT TAKE TO WRITE AN ARTICLE?

FIGURE 4.2 Number of articles (filled circles) and mean length of an article (unfilled squares) in the *Journal of Ecology* for each year during nine decades. Both variables corrected for the post-1991 change in format.

As Figure 4.4 shows, however, there are substantial differences in practice among different sciences, and these are not simply related to position along the ecology–biochemistry–physics axis. Figure 4.4 shows the frequency of authors per article in all four journals of the British Ecological Society for the single year 1997, and for *Physics Letters B* for the first half of 1997. They are fairly similar, and are different from the small crowds that were needed at that time to construct an article in Biochemistry. In both ecology and physics the commonest number of authors was two. In ecology there were slightly more three-author articles than single-author ones; in physics the reverse was true. Most of the single-author works in physics were about 'theory' – often a solitary activity. Three- to eight-author articles in ecology outnumbered those in physics. In this multi-author ecology group, articles about applied ecology, from government research stations, and originating outside the UK were over-represented. In 1997 in the ecology journals there were no articles with more than eight authors, whereas there were 40 (about 10% of the total) of the

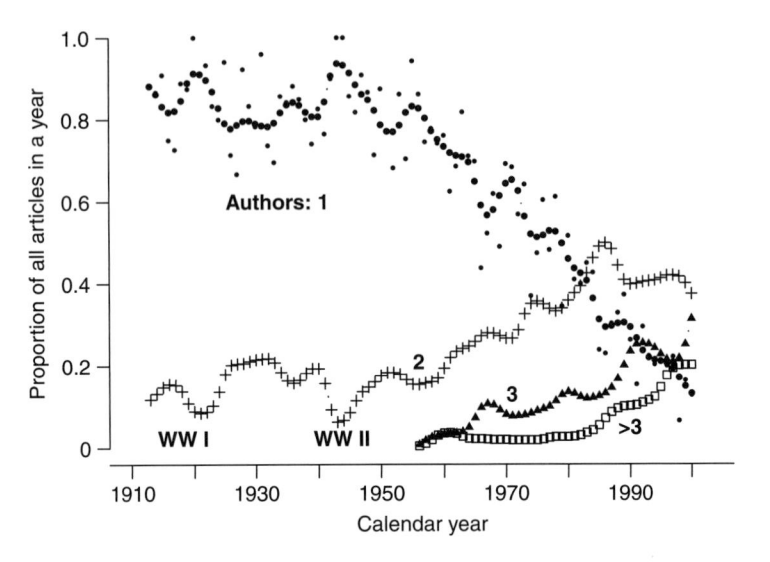

FIGURE 4.3 Proportion of articles with 1 (circles), 2 (crosses), 3 (triangles), and >3 (unfilled squares) authors (the maximum was 10) in the *Journal of Ecology* each year during nine decades. The values fluctuate erratically, as is shown in detail for single author articles (smaller filled circles), so for these and others I show only smoothed curves to make the trends more clear. The smoothing is '4253H' (Velleman & Hoaglin 1981). During the World Wars the number of single-author articles increased at the expense of dual-author ones.

physics articles that had anything up to 447 authors, as the upper part of Figure 4.4 shows. These were the productions of the necessarily large teams working on experimental particle physics. Many of these collaborations had or have their own group titles: 'Aleph', 'CDF', 'H1', and many others. Each group devises its own publications policy defining the circumstances in which an individual may be named as an author. These rules try, not entirely satisfactorily, to balance the conflicting requirements of group unanimity and individual credit to help career advancement (Galison 2003). The articles the group produces have to withstand internal examination more severe than most editorial and refereeing processes. Each member of the collaboration has an opportunity to see an article before it is published with their name on it, but their acquiescence in authorship is usually assumed

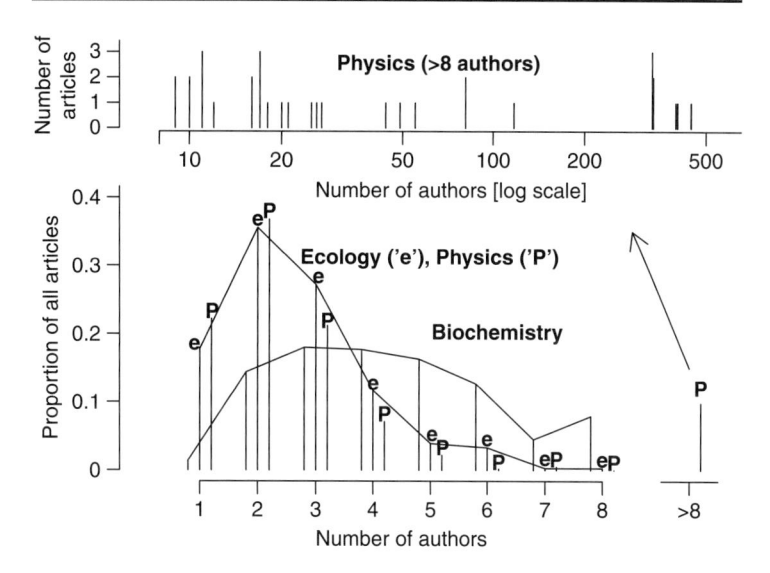

FIGURE 4.4 *Lower.* Proportion in 1997 of the total number of articles with 1 to 8, and >8, authors for three journals or groups of journals: 410 articles in the four journals of the British Ecological Society (*Journal of Ecology, Journal of Animal Ecology, Journal of applied Ecology,* and *Functional Ecology*); 355 articles in three of the eight volumes of *Journal of Biochemistry;* 360 articles in *Physics Letters B* during the first half of the year. Connecting lines to aid the eye are shown for Biochemistry and for Ecology. Physics is similar to Ecology except for the 40 physics articles with more than eight authors. The distribution of these is shown in the *upper* graph: note there the log scale of number of authors.

unless they specifically reject it. Some of the authors are experts in a field (such as high-speed electronics) essential to the team but not to the subject of the article itself. A few may not even understand a particular article of which they are an 'author'.

Large teams occur in molecular biology too: the mouse genome sequence was announced by 222 authors (Waterston *et al.* 2002) of whom the first four were leaders and the rest appeared in alphabetic order. The largest number of authors for a single article in the physics sample was 447, but this is not a record. The CDF (Collider Detector at Fermilab) group had 543 authors. The formal announcement of the properties of the Z° particle was made by 562 authors (Aarnio *et al.* 1989). In the medical field (Topol *et al.*

1993) 976 authors of an article described a medical trial involving 1081 hospitals. And a more recent particle physics consortium, the CERN ATLAS (A Toroidal Large hadron collider ApparatuS) has published an account of the apparatus with 2926 authors ranging from Aad to Zychacek from 169 institutions (Aad *et al.* 2008). The physicists names are usually in alphabetic order of family name, sometimes subsidiary to alphabetic order of contributing laboratory. Responsibility and credit in these circumstances cannot be inferred accurately from authorship (though private enquiry among members of the collaboration may reveal something).

Sometimes authorship is a gift, wrongly given for social reasons, or as a reward for persevering but routine paid assistance, or to aid promotion.

A different aspect of authorship comes into focus when one considers individuals. For example, the singular peripatetic mathematician Paul Erdős (1913–96) renounced personal possessions and pleasures, and lived solely for mathematics, moving from host to host, staying for a few days to work with the current host on a mathematical problem, then moving on to a different host. He was an author of about 1500 articles with a total of about 500 different co-authors (mostly in ones and twos). He was the main or, at least, an important contributor to most – probably all – of these articles. (Mathematicians may, light-heartedly, be allocated an 'Erdős number' which is the smallest number of other authors linked by joint authorship of a publication before one comes to an article with Erdős as an author. Thus if there is an article by Smith and Jones, and one by Jones and Robinson, and one by Robinson and Erdős, then Smith's Erdős number cannot be more than 3.)

Consider now the bibliography of Franz Halberg (undated), sent unsolicited to me some years ago. It records the fruits of a similarly long and apparently productive career in experimental science, much of it as a laboratory director: 1268 publications spread over 40 years. That is an average of 32 each year or one every 11.5 days, including weekends and holidays: an astonishing rate in experimental science. The majority of these publications have three to six authors. How much, and what, did Halberg contribute to them?

This is not an isolated case: the nature of the work and the culture of the subject in chemistry result not uncommonly in sustained publication rates for an individual 'author' (of multi-author works) of 30 a year or more.

You may think that the problems of large consortia, like those of the super-rich, do not affect you. But on a smaller scale any article by more than a single author does, potentially, create problems. Five decades ago most bioscientists worked alone or perhaps in pairs pursuing ideas that had come to them (Hubel 2009). Potential graduate students were expected to come with a topic that they wanted to work on, though this might turn out to be impracticable. They served an apprenticeship getting increasing independence. Now, however, most new graduate students will have a topic given to them. They may expect to collaborate closely with their supervisor to begin with, becoming more independent with experience and the passage of time. Increasingly the topic is a small part of a larger group project, and the group may have several postdoctoral members, and perhaps a dozen postgraduates. These changes have raised in an acute form the questions of who should be named as an author on an article, and what order should they be arranged in.

Who qualifies as an author?

One, over-simple and perhaps cynical, view of scientific authorship is that it is a currency or token the acquisition of which may create esteem for you, the author, which can be used to get a job, tenure, or promotion. The multi-author examples raise several questions. What constitutes authorship: how can one decide who in a small to medium-sized group of contributors should be named as an author and who should simply be thanked in the Acknowledgements (or not at all)? Should authors record which of them was responsible for which part(s) of the work? The potential credit to be had from authorship is obvious: what are the accompanying responsibilities – and risks? What happens if it later turns out that one of the authors has fabricated the data? Are the other authors to be held responsible too?

Does the order of authors generally imply anything at all? In the large physics collaborations it does not: Aaby always precedes Zyvoniev, at least within his or her own institution's list. In some parts of chemistry convention dictates alphabetic order. Some small groups of bioscientists that publish repeatedly together arrange to rotate the order of names in an arbitrary way. Practice is so varied in general that little reliance can be placed on the order of author's names, but to be first author *may* indicate a greater degree of responsibility and credit, and to be last *may* imply a supervisory or guarantor rôle. But in at least one field the last author is considered the most important. One fears that at least some members of promotions committees are unaware of this variability.

Other questions emerge in a particular sort of dual-authored article. In what circumstances should the supervisor of individual (not group) doctoral work appear as an author on work carried out by the student?

At one extreme the student had the idea, did the work, and wrote the article. All the supervisor did was to read and comment on it.

In the middle, the supervisor had the idea, taught the student some or all of the necessary methods, and suggested the analysis and perhaps the structure of the article, but all the rest was the student's work.

At the other extreme, the idea was the supervisor's, the experiments and methods were designed by the supervisor, the student did the work following instructions in the same way that a trainee technician would, and the supervisor analysed the results and wrote the article. (This is poor supervisory practice, but that is not the point here.)

Most people would probably say the supervisor should not be an author in the first example, should be second author in the middle one, and should be first (perhaps only) author in the last one.

Having the supervisor as an author implies some guarantee of the reliability of the work and, if the supervisor is well known, may get the article noticed more quickly.

As a student you need to keep in mind that the apparently simple idea that has proved so fruitful and that was put to you in ten minutes by your supervisor may have emerged from much previous work and thought by your supervisor (who is subject to periodic and worryingly frequent reviews of his or her publication record). As a supervisor you have to remember that most of the hard graft, and many of the detailed initiatives, were the work of the student who has no more than a single brief opportunity to establish a foothold on the slippery lower slopes of a scientific career.

What is clear is that it is wise for the student and supervisor to discuss authorship and the order of their names early in the work, and to keep the matter under review as the work proceeds and as articles are drafted, even though they do so in the shadow of the proverbial advice 'Don't count your chickens until the eggs have hatched'. Failing to consult on such matters can lead to friction and resentment, and even, sometimes, to the courts.

Another problem is raised by the increasing practice of adding a tail of parasitic hangers-on, sometimes known as 'guest-authors', or 'honorary authors', and ending with laboratory Director Cobley, T. ▼. What was the rôle of such lampreys and leeches? Are they in any real sense authors? If several attach themselves to an article which a student submits as part of a thesis, will examiners begin to wonder how much of the article is the student's work? Will the referees wonder why it took seven people to make such a simple experiment, and to ask how many of them are really authors? Some groups routinely name all members as 'author' (even when some have not contributed to the particular article at all) in the belief that it helps build team spirit. This seems to me to be a dubious and possibly dangerous practice.

▼ From the traditional English folk song 'Widdecombe Fair' which describes how Tom Pearce's grey mare is forced to pull to the fair an overcrowded cart containing "Bill Brewer, Jan Stewer, Peter Gurney, Peter Davey, Dan'l Widdon, 'arry 'awke, Old Uncle Tom Cobley and all". As a result "Tom Pearce's grey mare, she took sick and died."

A short article in *The Lancet* based on a study of 12 children suggested that there might be a link between MMR (measles, mumps, rubella) vaccination and the later development of autism. The article had 13 authors and became controversial, partly because of the implications, and later when doubts were raised about the reliability of the data. Ten of the authors later retracted the implications of the article they had 'authored', with some damage to their reputations, and the journal later retracted the whole article. This raises the question: who is responsible for the contents of a multi-author article?

Two-thirds of a sample of 2047 formally retracted articles were so because of overt or covert author 'misconduct', and the rate has grown at about 5% a year since 1974 (Fang, Steen & Casadevall 2012). In 2012 about one article was retracted for every 10 000 published. Whether we are seeing but the tip of an iceberg or all but the feet of a floating duck is unclear.

Some of these questions have recently become acute as a result of scientific frauds, especially, but not only, in the medical sciences (Gardner 1981, Broad & Wade 1982, Lock & Wells 1996, Park 2000). Scientific frauds are not new, and have occurred in many fields: the algal geneticist Moewus (Gowans 1976), the psychologist Burt (Rennie 1999), the biochemist Spector (Broad & Wade 1982), the medical biochemist Aberhalden (Deichmann & Müller-Hill 1998), and the physicist Schön (Kennedy 2002, Anonymous 2002, Reich 2009) are but notorious members of a much larger group. The perpetrator of the Piltdown Man fraud (passing off extensively modified ape skull bones as an early hominid of a form that agreed with then current expectations) is still unsure, though there are strong pointers.

Fraudulent actions range on a continuous scale upwards from the undeclared suppression of inconvenient outlying points on a graph, to deliberate small- or large-scale invention and falsification of data or even of whole experiments. To these frauds should be added the unacceptable practices of duplicate publication (Malmer 1997), and plagiarism, though these are misdemeanours rather than felonies, because they do not misrepresent Nature.

What drives some scientists to behave thus? Self-delusion, belief that one has a hotline to Nature (Herself, not the journal), seduction by a beautiful hypothesis (as is well known, a beautiful hypothesis may be slain by a single ugly fact), ambition and self-advancement, the tyranny of deadlines and of 'quality' assessment, tenure, and promotion procedures, covert and overt pressures from research grant givers (commercial and industrial, government, and charities), and the need to eat are common causes. See Berger & Ioannidis (2004) for lightly disguised, humorous, examples in the style of the Decameron. One sad example (Lock 1996) follows.

Dr Malcolm Pearce, an obstetrician and gynaecologist, published articles claiming in one to have brought an ectopic pregnancy to term (with the birth of a live child), and in another reporting a trial involving 191 women with polycystic ovary disease (a mercifully uncommon affliction). These articles were both published in *British Journal of Obstetrics and Gynaecology* of which Pearce was an assistant editor, and Professor Geoffrey Chamberlain (his Head of Department and, at the time, President of the Royal College of Obstetricians and Gynaecologists) was editor-in-chief. Chamberlain was also a guest-author on the ectopic pregnancy article, with which he had only a tenuous connection. Pearce later admitted that he had invented data in both articles and was struck off the register of doctors by the British Medical Association. Chamberlain resigned as President of the Royal College and as editor.

Science itself, though harmed by fraud, is self-correcting though the process may be slow, costly, inefficient (corrections rarely catch up with the original error), and may adversely affect the careers of those doing the checking. Results that cannot be repeated (be they the consequence of fraud, honest error, or chance) are rejected from the body of accepted knowledge. The more important the claimed result the more vigorous is the rejection. One must distinguish conscious fraud, of the kinds listed above, from the numerous instances of bad luck, honest mistakes, poor experimental technique, misinterpretation, confusion, ignorance of an unfamiliar but necessary field, or self-delusion (Close 1990, Gratzer 2000). Claims that the size of an effect is independent of the size of the cause, or reports of effects that are near the limit of detection are particularly suspect. Examples are: N-rays (thought to resemble X-rays, but really the result of subconscious bias, and rejection of 'unsatisfactory' results by the observers); polywater (the incorrect interpretation of the correct observation that in narrow silica tubes the viscosity of water is much larger than it is in bulk); water 'memory' (the claim, probably based on self-delusion, bias by a technician who knew what result was hoped for, and rejection of 'failed' experiments, that a solute still had effects even when diluted 106 orders of magnitude beyond the point where the average sample contained only a single molecule); cannibalistic memory (a report that a planarian 'worm' could eat another of the same species and then show behaviour that had been previously learned by the victim); the MMR vaccine scare (distorted evidence linking the triple MMR vaccine with the onset of autism); the initial claims, at least, of cold fusion (a claim that atomic fusion of hydrogen gas could be induced in the metal palladium at near room temperatures, the claim resulting from neglect of controls, calibration, background, and precision, and ignorance of the necessary physics, interacting with commercial and journalistic pressures); and many others. Several of these achieved initial notice mainly because they were the work of already-respected scientists. All were widely discussed in their time but, correctly or not, were not admitted to the body of established knowledge. Other instances, such as continental drift, were initially rejected on the unsatisfactory argument that 'because I can't imagine a mechanism therefore the hypothesis cannot be true', but were later admitted.

Like corpses left on gibbets, these examples all swing as jolting reminders of how slippery the path towards truth may be.

'Ghost-authors' are a rarer problem. A study was encouraged and supported by the tobacco industry (Hong & Bero 2002) but the industry involvement was partly hidden: at least two employees of a tobacco company, qualified by commonly used criteria to be authors, were not named as such. Both guest-authors and ghost-authors 'contributed' to trials of the Merck drug 'rofecoxib' (Ross *et al.* 2008). The extensive and systematic use of ghost-authors by the pharmaceutical company Wyeth is described in Fugh-Berman (2010).

Driven by examples of guest-authorship and ghost-authorship, and of fraud, the editors of medical journals stimulated discussions (Lock & Wells 1996) and began experiments with several reforms. An extreme suggestion is to abandon the concept of 'author' as beyond repair, and replace it with that of 'significant contributor'. A milder reform is to attempt to minimise what is seen as abuse by establishing criteria for authorship. The ICMJE (International Committee of Medical Journal Editors) has evolved rules which, as applied by a particular journal, state:

> "Authorship credit should be based *only* on substantial contributions to (1) conception and design, or analysis and interpretation of data; (2) drafting the article or revising it critically for important intellectual content; and on (3) final approval of the version to be published. All authors should meet conditions 1, 2, and 3. All persons designated as authors should qualify for authorship, and all those who qualify should be listed. Acquisition of funding, collection of data, or general supervision of the research group alone do not constitute authorship."

These rules are useful in as far as they exclude guest-authors and require the inclusion of ghost-authors.

You can see a more specific version of this approach in Table 4.1 while in Table 4.2, after Hunt (1991), is an approach amounting to 'significant contributor' in which each person's contribution to a piece of work is assessed numerically in five categories and anyone with a total above a threshold is offered authorship. The details in this scheme depend more on the nature of the work than do those in Table 4.1, so the Hunt scheme is likely to be of most use to large groups working on diverse projects but in a limited field. Nor was it easily applicable to an article (Grime *et al.* 1997) with 35 authors, 25 from the group usually using the Hunt scheme.

TABLE 4.1 Criteria for authorship, modified from Huth in Lock (1996) omitting specifically medical criteria

FEATURE	LEGITIMATE	**NOT** LEGITIMATE
Genesis of research	Development of testable hypothesis	Suggestion that (legitimate) authors work on the problem
Review	Critical interpretation of assembled articles and data	Suggestion that review be written
Research itself	Design of work Development of new method or critical modification of existing method	Suggestion of standard design or methods Routine observations or measurements
Interpretation	Explanatory insight into unexpected phenomenon	Routine reports
Writing	First draft or critically important later revision	Criticism of drafts and suggestions for revision of presentation
Responsibility for content	Ability to justify intellectually the conclusion of the article	Attesting to no more than accuracy of individual facts reported

A milder suggestion, now widespread among medical journals, is to name (as part of Acknowledgements) a 'guarantor' for the article overall, and to state what each author is responsible for. This strikes a nice balance: the specialists in the groups that are now necessary in many fields are not expected to be able to vouch for the work of other specialists (whose work they may not even understand), but the 'guarantor' will feel the need to take at least some trouble to assess the reliability of all the others.

Finally, in the problems of authorship, the journal *Nature* has moved to counter another emerging problem – the author with support from commercial sources – by asking authors to declare any financial or other support or financial interests in what they write about.

Journals themselves are not free of conflicting interests. An order for 900 000 (paper) reprints of one article putting a new drug in a favourable light contributed about 2.4 M$ to a medical journal's income, while about 40% of one medical journal's income came from reprint sales (Lundh, Hróbjartsson, & Gøtzsche 2012).

TABLE 4.2 Authorship scoring scheme used at the Unit of Comparative Plant Ecology for articles about experimental plant ecology, slightly modified from Hunt (1991). The contributions made by an individual are assessed and summed in five categories. The maximum for any individual is 20; anyone with a total of 5 or more qualifies for authorship

Intellectual input	Contribution to **data acquisition**
0 None	0 None
1 One detailed discussion	1 Small
2 Several detailed discussions	2 Moderate, indirect
3 Correspondence or longer meetings	3 Moderate, direct
4 Substantial liaison	4 Major, indirect
5 Closest possible involvement	5 Major, direct
Contribution to **data processing**	**Specialist** input
0 None	0 None
1 Minor or brief	1 Brief or routine advice
2 Substantial or prolonged	2 Specially-tailored help
	3 Whole basis of approach
Literary contribution	
0 None	
1 Editing others' contributions	
2 Small sections	
3 Moderate proportion	
4 More than half	
5 Virtually all	

Many disciplines are, sensibly, slow to change their traditional practices, but the mildest reform which requires a brief statement of responsibilities in Acknowledgements, thus justifying authorship, has much to recommend it. You can try this yourself even if it is not required: with luck the editor will accept it.

QUESTION:
For your previous or next article, who (according to the ICMJE criteria above) qualifies for authorship?

PART II
Improving

PART II

Improving

5

Style in writing

Writing style is recognisable by the choice and arrangement of words, and by the physical layout ('format') of the work. The style of an instruction manual is different from that of a novel or a play, and their styles differ from that used in a newspaper, in a legal document or in poetry. Scientific style is closely linked to the main objective of scientific writing: *to convey information as clearly and simply as possible.* ▾ Do this, and you may one day deserve the obituarist's note on the Nobel Prize winner Professor Sir Bernard Katz, whose first language was not English, that "His prose was simple, straightforward and unpretentious ... Every new entrant [to the field of neural physiology] should read his work from beginning to end."

> ▾ From the King James Bible, Ecclesiastes (9:11): "I returned, and saw under the sun, that the race is not to the swift, nor the battle to the strong, neither yet bread to the wise, nor yet riches to men of understanding, nor yet favour to men of skill; but time and chance happeneth to them all."
>
> Here is George Orwell's (1946) parody of this in a deliberately poor style: "Objective consideration of contemporary phenomena compels the conclusion that success or failure in competitive activities exhibits no tendency to be commensurate with innate

> capacity, but that a considerable element of the unpredictable must inevitably be taken into account."
>
> I read the first with pleasure partly for its simple, somewhat archaic and therefore fresh, language and the repeated strong images it creates, and I shuddered at the Orwell version.
>
> As an editor of a scientific journal, however, I would have to prefer a concise version of the same thought: "I found everywhere that rewards rarely reflect merit; chance rules all."
>
> (Passage above reproduced from *Politics and the English Language* by George Orwell. Copyright © George Orwell, 1946. Reprinted by permission of Bill Hamilton as the Literary Executor of the Estate of the Late Sonia Brownell Orwell. The whole article is worth reading.)

Style may be acquired by wide reading of the works of those generally held to be 'good' stylists, or by learning rules (and exceptions). Be aware, however, that many scientists and almost all editors hold strong views about style, and many – perhaps all – would disagree with at least some of what follows in this chapter.

Swift had a pithy view of good style: "proper words in proper places". (He used 'proper' in its older sense of suitable or appropriate.) More helpful perhaps is the advice of Matthew Arnold: that the secret of good writing lies in having something to say and saying it clearly. Gowers, revised Greenbaum & Whitcut (1986) quotes the 10-year-old schoolboy's essay on 'A bird and a beast'.

> "The bird that I am going to write about is the owl. The owl cannot see at all by day and at night is as blind as a bat.
>
> I do not know much about the owl, so I will go on to the beast which I am going to choose. It is the cow. The cow is a mammal. It has six sides – right, left, an upper and below. At the back it has a tail on which hangs a brush. With this it sends the flies away so that they do not fall into the milk. The head is for growing horns and so that the mouth can be somewhere. The horns are to butt with, and the mouth is to moo with. Under the cow hangs the milk. It is arranged for milking. When people milk, the milk comes and

there is never an end to the supply. How the cow does it I have not yet realised, but it makes more and more. The cow has a fine sense of smell; one can smell it far away. This is the reason for the fresh air in the country. The man cow is called an ox. It is not a mammal. The cow does not eat much, but what it eats it eats twice, so that it gets enough. When it is hungry it moos, and when it says nothing it is because its inside is all full up with grass."

(Passage reproduced from Gowers E, revised Greenbaum S, Whitcut J 1986 *The Complete Plain Words* 3rd edn. London, Her Majesty's Stationary Office. 288 pp.)

You may suspect that this was invented or at least 'improved' – would a 10-year-old have such command of the semicolon? – but the passage illustrates both the essence of a good style (saying something and saying it clearly) and the fact that style is independent of content.

Here, however, is an example of poor style from a serious report (by a government department) about universities:

"An overarching national improvement strategy will drive up quality and performance underpinned by specific plans for strategically significant areas, such as workforce and technology. The capital investment strategy will continue to renew and modernise further education establishments to create state of the art facilities."

This seems to mean "We shall think about what to do and spend money to make universities better" while in the following, quoted in Bulley (2006), the author seems to have been saying that "sometimes you feel you need to do something new":

"Problematic situations in which it is impossible to implement inherited repertoires of action may create reflective distance from inherited schemata and encourage or permit their transformation or substitution with other schemata available in an actor's environment."

Nearer home, the following is a fragment from a review, aimed at working scientists, of a book about the interrelationships of socio-economic factors and climate change:

"At the same time the important questions are preempted [sic] about whether such approaches ignore other socio-economic, political, cultural or environmental processes which influence climate changes and how these may affect local agriculture or world trade. The focus on the specific questions and uncertainties *within* that modelling paradigm aided by the *promise* of future improvement supplants any deeper questioning of the implications of the underlying epistemic commitment itself and of whether other kinds of approach altogether may be as illumining for policy. It is the promise, not the fact, of predictive validity via the (escalating) truth-machine identity which thus in important aspects becomes a policy heuristic which helps create an emergent global political order. Oreskes *et al.* have analysed some ways in which earth sciences modelling has had to make various commitments in the dark in order to produce apparent intellectual closure around particular model structures."

Further away but even more illuminating is the analysis by the physicists Sokal & Bricmont (1998) of the swirl of pretentious verbiage used by some authors to cloak the misuse of scientific concepts in fashionable French postmodern sociophilosophy. The website <www.elsewhere.org/pomo> (6 Sept. 2013) will construct plausible and fully referenced original essays in this content-free style for you, though without the misuse of scientific concepts. And the website <www.pdos.csail.mit.edu/scigen> (6 Sept. 2013) will 'write' a fully referenced nonsensical computer science article, with graphs and line diagrams, in the names of authors that you specify.

You may think that no scientist would ever write as obscurely as these examples, but the summary of an article in a well-known ecology journal tries, in five times as many words as in this sentence, to explain that moose often feed on plants in rivers and lakes but defecate and are killed by wolves on dry land thus moving nitrogen that was diffusely distributed in the water to concentrated patches on the land. Here are some of the phrases used: 'associated spatio-temporal patterns of resource flow across aquatic–terrestrial boundaries', 'cross-habitat resource flux', 'significantly clustered at multiple scales', 'faunal-mediated resource transfer'. The main text is just as glutinous.

The schoolboy's essay about a bird and a beast has few words of more than two syllables. Most hymns and carols use short words too. But the examples above of poor style have numerous three-syllable and four-syllable words. Gunning's light-hearted 'Fog Index' (F) is one of several 'readability' indices. In a modified version F is calculated for a sample of at least 100 words as $F = 0.4$ $W\,P$, where W is average words per sentence and P is percentage of words with three or more syllables (excluding words that are simple compounds such as 'minicab', words that begin with a capital letter, and words ending in '-es' or '-ed'). It is claimed that if $F > 12$ then an undergraduate student will struggle to understand the passage at first reading. To be widely comprehensible $F <= 8$.

The F value of the schoolboy's essay is 5; for the longer examples above F ranges from 33 to 39. The sentences are too long and too many of their words are polysyllabic. (And simple ideas are wrapped in too many words, some of them unusual.) For more substantial works, the mean of six similar indices for *On the Origin of Species* is 14, of the King James Bible 11, of *Pride and Prejudice* 10, of this book 9, of *War and Peace* (in English translation) and of *TheWind in the Willows* (a children's book) 8, of several books by Beatrix Potter (more children's books) and of James Joyce's *Ulysses* 6. This last is illuminating: *Ulysses* is difficult to understand. Simple words and short sentences are important elements of good style, but there is more, much more, to good style.

Here then is the essence of good scientific style: try to write clearly, precisely, and concisely in short sentences using simple words as much as possible, the words chosen so that your reader

is unaware of you as author and is never distracted by the form of what you have written.

If, at the end of the article, your reader's understanding has never been impeded and hir attention has never been diverted, then your style was good. My use of 'hir' in the previous sentence is poor scientific style because you were distracted into wondering:

is this a mistake, or is it an unfamiliar way to try to avoid 'his or her' or 'their' ?▼

> ▼ 'John and Jean knew *their* own minds' was criticised by Fowler & Fowler (1906) as bad grammar. The Fowlers' solution – ungallant as they admitted – was to use 'his' impersonally. A century later that has become unacceptable to many readers. 'Their' in this context is now widely used in speech. Caxton and Shakespeare *(Much Ado about Nothing 3:4)* used it in writing, and it is now so frequent in writing that even some pedants accept it (reluctantly). Nowadays, if I write that 'John and Jean each knew his own mind' you might suspect that Jean is French or Belgian.

The word 'style' is actually applied to text in three ways: first, for the arrangement of ideas in sentences, paragraphs, and larger sections; second, for the choice and arrangement of words within sentences; and third for details of punctuation, typography, and similar matters. Detailed advice about these details can be found in comprehensive style manuals such as Ritter (2002), Style Manual Committee, Council of Science Editors (2006), and Chicago University Press (2010).

The arrangement of ideas and words

'Aroccdnig to reearsch at an Elingsh uinervtisy, it deosn't mttaer in waht oredr the ltteers in a wrod are, the olny iprmoatnt tihng is taht the frist lteter and lsat ltteer are in the rghit pclae. The rset can be a taotl mses and you can siltl raed the txet whit few pormbels.' Thus is shown the dominance of the first and last letters in a word. The same applies to the first and last words in a sentence. One can go further: the last word or phrase in a sentence is the most prominent in the sentence, and the first word or phrase is the next most prominent. The same principle applies to paragraphs, though it will be sentences or ideas that are being placed. Typically something new, or a direct link to what is to follow, is put at the end and is thus given the prominence that allows easy flow to the next sentence or paragraph. For example the transition in '... further away from the delta I found more rushes. The delta is flooded ...' would be smoother as '... I found more rushes further away from the delta. The delta is flooded ...'. Instead of 'A ... B. C ... B' use 'A ... B. B ... C'.

'Proper words in proper places'

In a paragraph the reader works with, and you hope will remember, the *ideas*. In a sentence the reader's attention is on the meaning and interconnections of *words*.

Grammar and syntax are the rules by which a language shows the relation between words. Together they define an etiquette of good manners, respecting the needs of others. For a living language these rules are not absolute or constant: grammarians infer the rules from common usage. Just as commerce and administrators used to follow colonial invasion ('trade follows the flag'), so grammarians follow the language observing, recording, analysing, and codifying. To flout the current rules of grammar in scientific writing will irritate, and thus distract, at least some of your readers. Many of them were brought up under a different, sterner, régime and will be upset by egregious new uses. Scientific style should therefore use conservative grammar and syntax to avoid activating antique, often arbitrary, baseless, or plain wrong, beliefs. To frequently split infinitives or to end a sentence with a preposition may become acceptable grammar – Pinker (1994) explains why neither of these widely held beliefs has any merit – but courtesy to your older readers should prevent you doing these things more than occasionally.

Scientific style is no more unchanging than grammar is. Two centuries ago the *Naturphilosophen* (Natural Philosophers) were a group with strongly developed all-embracing theories about how Nature worked, but with little interest in observing how She actually functions (Hoffmann 1988). Scientific style evolved in the early nineteenth century to counter these views. It concentrated on facts presented in an impersonal way to minimise the writer's presence: it emphasised 'things' rather than 'people'. Reports were written with verbs in the passive – 'the solution *was boiled*' – suppressing the personal 'I' or 'we' element in what was pretended to be a passionless search for truth. Science in practice is not like that, however, and the customs of the time led to flaccid writing and awkward (and thus obtrusive) circumlocutions, such as calling oneself 'the present writer' instead of 'I'. This ritual persisted until a few decades ago, and some editors still encourage or require it. But the Natural Philosophers and their beliefs have faded into history. Kirkman (1980) asked the members of three large learned societies (Biochemistry, Ecology, Engineering) to judge six versions of a report of the same work (a different work for each of the three groups), each work written in six different styles. He discovered that practising scientists and engineers preferred to read accounts that were generally active rather than passive, with a judicious use of 'I' and 'we'. 'We' is less awkward and obtrusive than 'I', but the use of 'we' by sole authors still sounds pompous: such use is recognised only for monarchs, writers of editorials and mathematical proofs, the pregnant, and the nominated leader speaking for a group. During the last hundred years many scientists have ignored the old advice. In particular, responsibility for assumptions, decisions, or choices can thus be made plain (Webster 2003). Kirkman's finding, and the growing interest in science as a social activity, have encouraged a more vigorously active and personal style, though the passive has its uses (see later).

This changing view of scientific style also allows you to attempt to convey the sense of excitement you had while doing the work. Perhaps you have forgotten the astonishment you felt after weeks of tedious slog when the results first flashed on the computer screen and revealed a wonderfully clear pattern? Inflated adjectives – 'these superb results' – are not necessary but if the results really did astonish you, and you think they will astonish readers too, then some note of your personal reaction will go a long way to humanising your report (though this does draw attention away from the subject, and should be done rarely). 'The closeness with which the results in Figure X follow ... surprised me.' Editors may try to remove such personal reactions: they are probably misguided in doing so.

Tense – past or present – is another aspect of style that is changing and about which opinions and advice differ. Indeed the subject resembles the Sudd where the Nile is dispersed in dozens of anastomosing shallow channels, none of which can be described as the main or 'right' one. Tradition required the past tense for Methods at least, partly on the grounds that they were necessarily in the past. But there is conflicting advice about almost all the other parts of a standard article. A coexisting convention required that until your work had passed the scrutiny needed before publication it should be referred to in the past tense, but already published work, being now a part of established knowledge, attracted the present tense. Present (where appropriate) is often more direct and vigorous. Here, however, is a mixture that seems natural and manages to include elements of both views: 'Watson and Crick showed that the DNA molecule was a double helix. The repeat distance is 34 Å and ...'. All one can suggest is: try to avoid obtrusive inconsistencies of tense.

There are three sorts of grammatical 'rule'. (1) Structural rules, such as those describing word order ('Experiment started big a today I'). The result of breaking a structural rule is likely to appear to a reader as nonsense. But Chomsky's famous illustration 'Colorless green ideas sleep furiously' does not break structural rules: it is the meaning of juxtaposed words that makes no sense. (2) Rules whose breaking reveals language that is not standard in scientific communication though it is in its own environment. Outside my window in the street I have just heard: "She be here." "You seen her coming?" "I done see her coming." (3) Conventions, many of them optional or plain wrong, and *bêtes noires:* split infinitives, 'hopefully', 'very unique', 'between the three', 'under the circumstances' and an ever-changing kaleidoscope of similar words and phrases.

You break the first two of these 'rules' at your peril. Good scientific style seeks to avoid distraction by observing even the third kind – most of the time.

Guidelines

(i) Aim to be brief, simple, clear, and vigorous. Try to write short sentences. Avoid multiple parenthetic remarks. Here is a fragment from the Translator's Note in a recently published translation of a classic German work. "Secondly, X frequently writes very long sentences, with diversions set off between commas, and parenthetic qualifiers, and even qualifiers of qualifiers, also set off between commas, and such constructions, while commonplace in the nineteenth century, do impede understanding, particularly by those raised on modern scientific English style. The previous sentence may give a flavour of the problem." Such a long sentence should be split into two or several sentences, usually at conjunctions such as 'and', 'but', 'if', 'also'. You will not always succeed in keeping sentences short and the longer ones will give variety. Every bioscientist (even biochemists) should read *On The Origin of Species*, but in it Darwin wrote sentences that average 40 words, and one of which was 208 words long. Do not emulate him in this. A common pattern is a short sentence to introduce the topic of a paragraph, then longer ones to develop the theme, and a final short sentence. The paragraph just ending shows this.

(ii) Prefer short words to longer ones, 'so' to 'consequently'; 'raise' to 'elevate', 'got' to 'obtained' and 'procured', 'pick' to 'select', 'saw' to 'observed'.

(iii) Often, but not always, prefer the active to the passive (provided that it does not lead to too frequent use of 'I' or 'we'): not 'measurements were made ...' but 'we measured ...'. Allow three exceptions.

First, when a second sentence refers to a subject introduced in the previous one. 'Great advances in cosmology followed the discovery of the red shift of the colour of light that results from the velocity of separation. The Doppler effect explains the red shift.'

You do not discover that the second sentence is still about the red shift until you get to its end. It is more easily followed if 'the red shift' links the sentences and the second sentence is in the passive: 'Great advances in cosmology followed the discovery of the red shift of the colour of light that results from the velocity of separation. The red shift can be explained by the Doppler effect.'

Second, if the subject is long. 'A great storm resulting in obstructed access to the site and in failure of the electricity supply ruined the first experiment' keeps your reader waiting too long to discover that the experiment was lost. The passive 'The first experiment was ruined by a great storm that . . .' is easier to follow. But make sure that both the subject and the (passive) verb are near the start of the sentence, as they are in the example but are not in 'Four replicates of plots with six species in each of two sites with contrasting climate and geology were established.' This sentence would, in any case, be better in the active form 'We established . . .'.

Third, to suppress the personal 'I' or 'we' in, for example Methods: 'the mice were kept in . . .' rather than 'we kept the mice in . . .'.

(iv) Prefer the specific to the more general. The police may say that 'a young male approached the premises', but you should say that 'a boy went to the shop'.

(v) Be sparing with adjectives and adverbs. Use adjectives when they are essential to specify type or kind ('green plants') but be wary of vague adjectives and adverbs such as 'very', 'small', 'large', 'rapidly' where a precise numerical value could be given, and of unreferenced 'comparatively', 'relatively' (in both cases 'to what?').

(vi) Generally, put evidence before your inference: not '. . . these dead cells contained no cytoplasm' but '. . . these cells contained no cytoplasm and so were dead'.

(vii) Avoid obscure words and circumlocutions: call a spade a 'spade' not a 'geotome'; 'jump to conclusions' if you must but do not 'make precipitate judgements on the basis of first impressions'. Replace 'showed an increase in magnitude with proximity to the boundary' with 'were larger near the edge'.

(viii) Jargon – the sort of technical expressions and abbreviations understood and used by everybody in your field – can save a lot of space and time and is acceptable in limited amounts as long as it does not impede understanding. Here is an example of the proper use of jargon (and a mistake) from the blurb for a mathematics book titled Sub Subgroups of Groups:

> "A non-trivial group is either simple or contains a proper, non-trivial, and therefore sub, subgroup. [This book] gives an account of sub subgroups of both finite and infinite groups, from the early work of Wielandt ... to recent results relating to the elusive submoralizer [sic] of a subgroup."

(ix) Avoid 'elegant variation' and periphrasis (wrapping an idea in an oblique, often elegant but obscure, expression such as 'I was economical with the truth'). There is a belief, common among journalists, that one should not use the same word (of five or more letters) twice in the same or neighbouring sentences even though one intends the same thing. In a scientific report this practice leaves readers wondering if the change from 'case' to 'patient' to 'child' to 'participant' to 'respondent' to 'person' to 'subject' implies some subtle difference and eventually concluding, irritably, that it does not. 'There were six positive cases but not one of these instances could be relied on' is poor scientific style (and 'instances' is redundant anyway).

(x) But also avoid using the same or a similar word with different meanings in the same or adjacent sentences. 'He will present the present presently' will cause difficulty for readers – even those whose first language is English. The chairman of a music company who wrote that 'The company has suffered unprecedented thefts of its [vinyl] disks, so I am forced to record record record losses' was not helping his readers combining as he has both this and the previous fault in a single sentence. *Roget's Thesaurus* (Roget revised Dutch 1962) is a useful source of synonyms (words with nearly the same meaning). It may be useful to scientific writers who wish to avoid using the same word with different meanings, but should not be used to avoid using the same word with the same meaning.

(xi) Avoid long adjectival phrases, especially those in which the qualifiers come before the thing they qualify ('stacked modifiers', colloquially 'goods train' or 'freight train' adjectives). For example, 'wheat pollen mother cell chromosome defect', which builds a stack of five modifiers before telling us what is to be modified, and is more comprehensible as a 'defect in a chromosome in the pollen mother cell of wheat'. The revised version is a bit longer but is more easily understood, and also encourages you to say which chromosome the defect is in ▾. Even worse is 'wind turbine design engineer's office move timetable' which creates images first of wind turbines, then of their designers, then of the designers' office, then that they are moving, and only after all these false starts to the fact that here we have a timetable for that move. Again, avoid the string of pre-qualifiers of 'timetable' by 'timetable for the move of the wind turbine engineers' office' or even 'timetable for the move of the office of the wind turbine engineers'.

> ▾ The structure of sentences and their parts is loosely parallel to calculation algorithms. Compare the Japanese form 'man dog bites' with the English 'man bites dog'. In the first the subject and object ('man', 'dog') are assembled, then the operation linking them ('bites') is effected. This is how you create a piece of furniture from a flatpack, and is how the algorithm for arithmetic calculation in computers and in early Hewlett-Packard calculators works (3, 9, +). It is known as 'postfix' or 'reverse Polish notation' (RPN) after the mathematician J. Lukasiewicz who codified calculation formalities. Alternatives, less useful for computation, are 'infix' (3 + 9) equivalent to ordinary English construction and modern hand calculators, and 'prefix' (+ 3, 9) notation. To stack modifiers is similar to RPN in that you assemble the modifiers before announcing what they modify. Pinker (1994) discusses these and other sorts of structure.

(xii) Be wary of repeating syllables or sounds ('selecting while anticipating preventing', 'found around the ground') or similar but unnecessary words ('follow the following steps': replace by 'follow these steps:').

(xiii) Short adjectival phrases are not always avoidable. The order of multiple adjectives before their noun usually puts the most important adjective next to the noun, while importance diminishes with distance from the noun. Compare 'the former Italian President' (not the present Italian President) with 'the Italian former President' (not the former President who is Belgian). Use a hyphen to show the affinity of a potentially ambiguous adjective: 'a small-scale model' is not the same as 'a small scale-model', and 'an Indian art-collector' is a native of India who collects art, but 'an Indian-art collector' is a collector of Indian art. Samples taken at '24-hour intervals' were taken once each day, those taken 'at 24 hour intervals' were probably taken at hourly intervals, but the phrase is ambiguous and would be clearer as '24 samples were taken at consecutive hourly intervals'.

(xiv) Experiment with the order of multiple qualifiers. A useful starting order is Opinion > Size > Age > Shape > Colour > Origin > Material > Purpose. For example '[An] interesting small new square red Welsh wooden educational [toy].' 'Ephemeral mosses first appear in frershly ploughed fields in autumn' creates a mental picture of short-lived mosses in ploughed fields, while 'Ephemeral mosses first appear in autumn in freshly ploughed fields' emphasises the time of moss appearance.

(xv) Be alert for ambiguity. Here are a few examples.

'Then, Beckham hinted, players began to hit the ball long after losing faith in the team's creative players.' Hit the ball long? Or hit it long after?

'I left the field early as it was getting dark': because you could no longer see, or it happened to be getting dark when you left?

'The threatened bryophyte database ...': is it the bryophytes (mosses and liverworts) or the database that is threatened?

The assurance, concerning a Degree Ceremony, that 'you will be asked whether you wish to attend before your final examination' actually meant 'you will be asked before your final examination whether or not you wish to attend'.

A newspaper reported that 'Senator X told Y that he was not the first to cheat Indians': is 'he' X or Y; was this an admission by X or was he accusing Y? The singular pronouns ('he', 'she', and 'it') are probably the most frequent cause of ambiguity.

'Three hundred authors signed off on the article'; did 300 agree to authorship, or did they refuse it? And what about the confusing adjacent 'off on'?

'The complaint was that she pulled the trigger that killed the moose without a licence.' Supplementary qualifications ('without a licence') try to attach themselves to the most recent plausibly relevant subject ('moose') rather than the more distant one intended ('she').

Test each sentence as you go along for possible ambiguity. Would it be possible for a reader to misunderstand your meaning? Do you need another comma? The qualification you have just written will try to attach itself to the immediately preceding item. Is that what you intended, or is this a 'moose without a licence'? Then put the text aside for a few days and re-read it seeking ambiguities.

(xvi) Lastly, here are a few more-detailed points.

1. Operations and procedures are applied to an individual, then repeated on numerous others. Similarly, the numerical values recorded in a table are individual though the table may contain many such numerical values: 'Values are concentration of ...' rather than '... are concentrations of ...'.

2. The choice between 'a' or 'an' is usually determined by how the words sound: 'an aardvark', 'a zebra', 'a hydrogen atom', but 'added an H ['aitch', the letter] to give ...'. There are a few difficulties: 'an NERC grant' assumes that NERC is an abbreviation in which the letters are separate ('an N-E-R-C ...'), while 'a NERC grant' assumes that NERC is an acronym pronounced as a single word to rhyme with 'jerk'.

3. Avoid beginning a sentence with an abbreviation, a number larger than ten (which would be shown as digits such as '127'), a person's name (particularly a citation), or with a lowercase letter (as in 'pH') particularly if the previous sentence ended

with another in this list, or anything other than a standard word with a lowercase initial letter. Your reader probably misses the full stop, runs into grammatical difficulty, and has to go back to reread more slowly and carefully. Not '... has been shown (Smith 1988). pH was ...' but '... Smith (1988) showed this. The pH was ...'.

4. Try rewriting one of your completed paragraphs to reduce its length. It may be helpful to practice in another field. As an example, here is the introduction to an interesting article (Wieser-Jeunesse, Matt, & De Cian 1998) in a respected chemistry journal. You need to know that a calix[4]arene has four upright equally spaced six-carbon rings that point slightly outwards resembling a crown or chalice (hence the name 'calix') and thus enclose a conical hollow. The 'phenoxy' group has an oxygen atom attached to one of a ring of six carbons. You also need to know that chemists are excited, for good reasons, by the chemistry of organic molecules that incorporate one or more metal atoms. Here is the published original.

> "A major attraction of cone-shaped calix(4)arenes concerns the presence of a macrocyclic cavity defined by four symmetrically sited phenoxy rings. To date, exploitation of such organised structures has mostly relied on converging π systems on the inner side that facilitate weak binding of various substrates, including certain metal cations. Surprisingly, despite increasing interest in the application of calixarenes as ligands in transition metal chemistry, the interior of the cavity has not been used to entrap or confine reactive fragments bound to transition metal ions. Such architectures could possess the capability to promote metal-centered reactions that are sterically constrained, thereby allowing combined shape control and regioselectivity. Furthermore, it is likely that the cavity walls will afford protection of highly reactive "M-R" units against undesired side reactions. We now report the first calix[4]

arenes with organometallic fragments positioned inside the larger opening of the cavity."

(Passage reproduced from Wieser-Jeunesse C, Matt D, De Cian A 1998 Directed positioning of organometallic fragments inside a calix(4)arene cavity. *Angewandte Chemie International Edition English* **37**: 2861–2864. Reprinted with permission by Wiley-VCH.)

To a chemist this is all easily understood. Could the main points be put more simply for a general audience? Here is my attempt that is 1/3 shorter than the original. Jargon that can scarcely be avoided is underlined.

'The large conical cavity of calix[4]arenes, enclosed as it is by four nearly vertical but outward pointing phenoxy rings with π systems on the inner side, can contain, bind with, and shelter other molecules. A few compounds with metals trapped weakly in this way have been made, but we now report the formation of compounds that have trapped organic fragments bound to transition metal ions. These complexes may allow control of the type and specificity of reactions between highly reactive trapped molecules, which are forced by the enclosing crown into particular configurations and protected by it from unwanted side reactions.'

Most scientists can tell when an article is well written even if they cannot pin down quite what makes it so. When you find such an article do spend time analysing what has made it a pleasure to read.

Punctuation and typographic details: the journal's rules

Style in its third sense concerns details of punctuation, typography, formatting, and similar matters. Opinions differ about most of these, and many details are determined by the style rules of the journal.

Punctuation

A detailed account of punctuation is beyond the scope of this book. Here I give only a brief account of a few of the more common uses. A more detailed, and entertaining, account is in Truss (2003).

Lawyers minimise punctuation to reduce the possibility of ambiguity, and advice is often given to follow this lead. But nearly everyone ignores the advice and uses punctuation for the benefit of readers: to improve clarity and readability. Which of the following variants do you prefer?

'... the most important things to him were cooking his books and his family'

'... the most important things to him were cooking, his books and his family'

'... the most important things to him were cooking, his books, and his family'

'... the most important things to him were his family, his books, and cooking'

(The third and fourth versions use the 'Oxford comma' – see later.)

The power of punctuation used in this way can be seen in the sentence

'Johnny where Jimmy had had had had had had had had had had had the teacher's approval'.

This incomprehensible string of 11 'had's is transformed by punctuation to

'Johnny, where Jimmy had had 'had', had had 'had had'. 'Had had' had had the teacher's approval'.

Another common piece of advice is that punctuation should not change the meaning, but (again) this advice is often ignored. Compare:

'She stopped crying' [she had been crying but then stopped doing so] with

'She stopped, crying' [she had been walking but then stopped and was crying].

Punctuation that affects structure: comma, semicolon, colon, and stop

The punctuation marks that affect structure are comma, semicolon, colon, and stop, where 'stop' represents full stop (period, point), question mark, and exclamation mark. The exclamation mark is rarely needed or acceptable in scientific text.

These marks serve two main purposes: to make the relationship between parts of a sentence or paragraph clear, and to express feeling. Fowler & Fowler (1906) call these logic and rhetoric.

Commas can alter the logical meaning: recall '...she stopped, crying'. Less immediately obvious are the following.

'...ants and bats, or cats ...' means (A + B) or C.

'...ants, and bats or cats ...' means A + (B or C).

'...ants, and bats, or cats ...' and '...ants and bats or cats ...' are both unclear.

Rhetorical effects can be easily fine-tuned by punctuation.

'The situation is perilous but there is still one chance of escape.'

'The situation is perilous, but there is still one chance of escape.'

'The situation is perilous. But there is still one chance of escape.'

The logic of these three is identical, but their rhetoric differs. They indicate different depths between the first and second thoughts. In the first example one has barely recognised the seriousness of the position before hope is offered. In the second, recommended in Strunk & White (2000), there is a perceptible pause though not long enough for despair. The third version allows the reader's spirits to sink deep before being galvanised by hope.

Whether or not to put a comma after the penultimate (last but one) item in a list is, like the 'big-endian' heresy in *Gulliver's Travels*, a largely arbitrary matter for the individual to decide. The Oxford University Press style (Ritter 2002) does require a comma in this position, so it is called 'the Oxford comma'. For example, compare '... pitching, yawing, and rolling are the movements on the three main axes.' with '... pitching, yawing and rolling are the ...'. The first is clearly A + B + C, while the second might imply A + (B + C). The logic is the same but the readability may be better with the Oxford comma. In ' ... 1950s articles by Kendrew, Watson and Crick and Perutz ...' the two 'and's cause confusion that can be removed by '... articles by Kendrew, Watson and Crick, and Perutz ...'. But in '... the structure of proteins, and RNA and DNA ...' the reader can do without a second comma. Some individuals are passionately in favour of, or against, the Oxford comma and advocate its consistent use or avoidance. This seems to me to be too rigid: consider your readers and if you think it will help them then use a comma in this position; if not then omit it.

The comma may be used if one or more items are possible: 'My cousin Jim is ...' (he is one of two or more cousins) or 'My cousin, Jim, is.. ' (my only cousin who, you may like to know, is called Jim).

Matter that follows a semicolon usually adds in an unpredictable way to what went before the semicolon. 'Man proposes; woman takes a while to make up her mind.' In general, what follows a semicolon should be able to stand alone as a grammatically acceptable sentence with a verb.

After a colon you may find an itemised or illustrative list, or a restatement of, or a striking contrast to, what went before the colon. 'The most used punctuation marks are the structure-affecting ones: comma, semicolon, colon, stop', 'Man proposes: God disposes.' What follows a colon, if not a list, should be a grammatically acceptable sentence with a verb.

Punctuation may be used in a hierarchic series: comma < semicolon < colon < stop can indicate breaks of increasing vehemence or of different kinds. Compare the following two versions:

> 'Philosophers study ideas, mathematicians numbers and logic, physicists and chemists the inanimate world, and bioscientists the living world; these definitions are often too rigid'

> 'Philosophers study ideas; mathematicians numbers and logic; physicists and chemists the inanimate world; and bioscientists the living world: these definitions are often too rigid'

In the first of these the four elements in the list are separated by commas, and the final caveat (warning) by a semicolon – the next higher-ranking mark. The commas are logically necessary (without them the sentence makes little sense) but the semicolon serves to emphasise that the caveat (a rhetorical device) is different in kind from the preceding items in the list. The second version moves both punctuation marks up a rank, thus increasing the emphasis of both the list and the caveat, but keeping the distinction between them.

If we needed to use a comma to extend one of the list items, giving elements at three levels, then we would be forced to use the second form:

> 'Philosophers study ideas; mathematicians numbers, logic, and inter-relationships; astronomers, physicists, and chemists the inanimate world; and bioscientists the living world: these definitions are often too rigid'

In these examples the semicolon and colon are used simply as breaks, of differing weight, to reveal the logical structure.

Comma, semicolon, colon, and stop have to satisfy numerous needs both logical and rhetorical, so there are many special 'rules' beyond the scope of this book. Here are three guides: (i) Use only as much punctuation as is necessary to make the text easily understood. (ii) Punctuation is to help your reader, not to make your life easier. (iii) If 'rules' conflict put logic before rhetoric.

The apostrophe, hyphen (and various dashes), and inverted commas are non-structural punctuation marks. The first two need separate consideration.

The apostrophe

The uses of the apostrophe are governed by complicated 'rules' distilled from common practices. Here are the four commonest.

To avoid confusion in this section I use double inverted commas (quotes) to enclose examples that use single apostrophes.

1. An apostrophe is *not* needed for a simple plural: 'Raining cats and dogs, hailing taxis.' The misuse of the apostrophe for a plural is sometimes called "the greengrocer's apostrophe" or, for the collective, "the greengrocers' apostrophe" after the common advertisement of "ripe banana's".

2. The apostrophe can signal missing characters as in "It's a long way to Tipperary" (for "It is a long way ..."), "... the early '60s" (the context to explain the missing century). The decade is "the 1980s" but is often "1980's" in the USA. "They'd 've been late ..." is an extreme example useful mainly to mimic speech. Note that "shall not" becomes "shan't" (not "sha'n't"), "they are" becomes "they're", and "will not" becomes "won't". In scientific writing constructions such as "can't", "couldn't", "shouldn't", "wasn't", "don't" usually seem too colloquial and are best avoided.

    Some words once required an apostrophe but no longer do: "omnibus" became " 'bus" and is now "bus"; "influenza" progressed through " 'flu" to "flu", photograph through "photo' ", and telephone through " 'phone". ("Refrigerator" acquired a 'd' to become "fridge", omitting the "'frige" stage.

3. The apostrophe can show possession as in "The individual freeman's right to vote". If the possessor ends in 's' and is a single syllable then usage is the same, as in "St James's St", "Strauss's waltzes". But if the possessor ends in 's' and has more than one syllable then the terminal 's' is usually omitted as in "The barons' rights", "Archimedes' principle". Some unusual plurals do not follow this pattern though: "women's rights" rather than "womens' rights", and "children's" too.

4. The apostrophe itself is omitted, however, in the possessive pronouns 'hers', 'its', 'ours', 'yours' and 'theirs'. "Mind its bite: it's dangerous" shows both the third and second uses.

Hyphens and dashes

A hyphen is '-'; a minus is '−', an 'en-rule' or 'en-dash' is '–' (about the same width as an 'N'); a still longer dash '—' is an 'em-rule' or long dash (about the same width as an 'M' and often the same length as the font is tall). Here they are, separated by spaces: '- – — −'. Notice that the minus and en-rule are similar but not identical. A hyphen is usually used to link a qualifier to what it qualifies: 'able-bodied', 'hard-boiled', 'fast-moving'. Do *not* use a hyphen for the minus sign: a '−' is available in the 'mathematical symbols' section of the 'special character' option in word-processors. Or you may be able to persuade your word-processor to let you specify then convert Unicode 2212 ('u2212', see Special Characters below) for the minus sign.

- An 'en-rule' (u2013) without enclosing spaces is used to show a range ('A–Z') or to link two words of equal status such as 'East–West relations' (in which you might logically reverse the positions of the linked items). The 'en-rule' may also be used with enclosing spaces, if that is the publisher's convention, to separate phrases or clauses. You may be able to set your word-processor to convert two successive hyphens into an 'en' rule. For most purposes you may use either the 'minus' or 'en-rule'. Try to keep to the same one throughout a document.
- An 'em' rule (u2014) with enclosing spaces may be used to separate phrases or clauses.

Word-processors treat a hyphen as if it were a place after which a line may be broken. You can usually get the word-processor to use a 'hard hyphen' that appears on the screen and in print to be no different from its normal 'soft' equivalent, but which is treated as an alphanumeric character would be and thus prevents the 'word' it is in being broken between lines. There is a 'hard space' that behaves in the same way. These 'hard' characters are particularly useful in preventing complex units of variables being broken between the end of one line and beginning of the next.

The hyphen '-' has many uses; use it only when essential, but do use it when it is. (A phrase such as 'iron-containing sands' can be written as 'sands containing iron'.) The hyphenation rules devised by publishers and word-processor designers are numerous, idiosyncratic, and depend on the printed column width. Switch off the automatic hyphenation option in your word-processor and leave the editor or publisher to apply their line-breaking hyphenation rules if they want to.

Compound nouns, especially technical ones, often begin as hyphenated words, but there is a tendency to join them as a single word: 'word-processor' is tending to become 'wordprocessor'.

Where a noun is preceded by a single qualifier that has no tendency to join the main noun a hyphen is unnecessary: 'lion cage'. A hyphen is usually used in examples such as 'twenty-one' and in 'south-east' (the direction), but not in 'Southeast' (the place). A hyphen is sometimes necessary to make the meaning clear: the inhabitants of the Australian state of Victoria would have responded with joy to the news that the Kelly Gang had 'reformed', but with fear had it 're-formed'.

Enclosures

Printers have specific names for enclosures: ' ' and " " are single and double quotes or inverted commas; '()' are parentheses; '[]' are brackets or square brackets; '{}' are braces; '< >' are angle brackets; and '-' and '—' are 'en' and 'em' rules (again). Enclosures should be paired, but the second '-' or '— ' is omitted if it is at the end of a sentence – as this one would have been. (The terminating ' – had the sentence continued.' has been omitted.) The publisher may specify the order if more than one of parentheses, brackets, and braces is to be used in equations, but if not then use the order: {[(...)]}. In ordinary text use only (...) and avoid nested parentheses such as those that would result from an author–date citation inside a parenthetic note as in: '... unlikely (because the method is too insensitive (Brown 1993)) ...'. Recast this without one set of parentheses as '... unlikely because the method is too insensitive (Brown 1993) ...'. Angle brackets have become common surrounding universal resource location (URL: website link) addresses. Angle brackets are used individually also, of course, as the 'less than' and 'more than' symbols.

When quoting it is sometimes necessary to insert a word that is not part of a quotation in order to make the quotation understandable out of context. Such inserted words should be enclosed in brackets, '[]'. Omitted matter is shown by an ellipsis '...' which has exactly three dots separated by spaces. For example: "He [Jones] said that he ... did not see the event."

Enclosures or containers for short diversions from the main line of thought in a sentence are, in order of increasing distinctiveness, commas < parentheses < dashes. (Some writers reverse the order of dashes and parentheses.) I might have flagged a diversion in the following increasingly emphatic ways:

'are, in order of increasing distinctiveness, comma, ...',
'are (in order of increasing distinctiveness) comma, ...'.
'are – in order of increasing distinctiveness – comma, ...'

Footnotes, endnotes, and other jumps

Footnotes and endnotes are where parenthetic comments, side issues, technical details, and (but not usually in scientific journals) references can be tucked away. Footnotes appear at the bottom of the page, endnotes are collected at the end of the article, chapter, or book. If the material in footnotes (references for example) is exclusively the sort of thing that almost all readers will want to skip, on first reading at least, then the footnote is a good way of clearing the undergrowth. But if the footnotes contain a mixture of sorts of material, some of which is for reference but some is interesting though parenthetic, then the reader has to divert to each footnote 'just in case' (or interrupt to read the footnotes out of context at the page end). Mixed sorts of endnotes are even worse as the diversions require page turning. Perhaps this is why footnotes and endnotes are discouraged in scientific reports, except perhaps on the title page, and in *Science*, *Nature*, and some other journals which use mixed endnotes instead of Methods and References sections.

References by page number to other parts of your article are even more strongly discouraged. In a journal that publishes in parts containing several articles the page references cannot be completed until page numbers are known, after all the preceding articles have been proof-read and corrected. Modern publishing programs help, but the process is still error-prone.

Individual publishers have their own standard typographic symbols and order for footnote or endnote cues (also called anchors) in the text. Here is a footnote.[@] I have used '@' as the cue to the footnote, but that is unusual. One common set of cues is asterisk, dagger, double-ended dagger, section mark, paragraph mark, double bar; followed by the same sequence of doubled characters: '*', '†', '‡', '§', '¶', '‖', '**', '††', '‡‡'. The '*' and '**' have other uses too and the journal may exclude them from its list. The dagger '†' is often used to show that a named person is dead. It may be best to avoid it for other uses if you can. This system works well if footnotes are rare, but if the cues refer to endnotes, or are frequent (as they may be on the title page of an article where the affiliation of numerous authors must be given in footnotes), then they are best cued by sequential number. Make sure that the cue cannot be misunderstood as a qualifier or power or index. Confusion may be unavoidable if citations to references are by superscript numbers. Special character cues may be useful though in tables set in text that has number citations.

Prefixes to people's names

The formal conventions are not simple and depend on the nationality of the name bearer and the language of publication. The main points are these. The French prefix 'de'; Dutch 'van', 'van der', and 'van den'; and German 'von', 'von der', and 'vom' (and many others) all mean 'of' or 'from' and all should have a lowercase

---

[@] A footnote. What do you think of it? It is easy enough to move to the footnote – you usually know fairly exactly where it will be – but getting back to the place you diverted from requires that you have remembered or marked where you came from, or that you make a distracting search. This is particularly bad when the footnote spreads onto, or even begins on a subsequent page. An alternative is to leave footnote-worthy material in the main text but to print it in a distinctive, usually smaller, type face. In this book I experiment with lightly shaded 'panel-notes' for the same purpose. They are easier to skip than small type, are easier to read (being in the same size type as the main text), and the place to return to, if you do divert to read them, is no more than the preceding paragraph away. This footnote might by chance be spread over two pages too, making the search more difficult. Now, search for the cue from which you were directed.

initial, unless the name bearer usually uses a capital initial (in Belgium 'De' is for commoners, 'de' for the nobility), or the word is at the start of a sentence (where prefixes are better avoided). In lists, convention requires that alphabetic position should be determined by the name that the prefix qualifies. But few writers and fewer readers are aware of this so in practice the prefix itself is often given an initial capital and the whole name listed under 'D' or 'V' as appropriate. Expect to find the Dutch author Hertz van Rental under 'van Rental' or 'Van Rental', not 'Rental, H. van'. A detailed list of conventions is in Style Manual Committee, Council of Science Editors (2006).

Abbreviations and italicised words

The simple rule is that an abbreviation has a terminal '.' only if the last letter of the abbreviation is not that of the full word. For example Mrs, Dr, Prof., ed (edited), ed. (editor), eds (editors), St (Saint, street), Rd, Av., Ave, *et al.*, Fig., etc. The full stop (period, point, or even, in email addresses, dot) at the end of the previous sentence is merged, as it is not at the end of this sentence, with that for the 'etc.'. There are many exceptions, for example units of variables (kg, cm), mathematical functions (such as abs, cos, exp, ln, log, mod, sin, tan), and accepted abbreviations (such as DNA, ATP) and acronyms in which the abbreviation is spoken as a word (AIDS, CERN, NATO). 'NERC' and a few other examples may be treated as an acronym ('NERC' pronounced as in 'work') or as an abbreviation ('N-E-R-C').

A general rule is that Latin (and other foreign language) words, or abbreviations of them, are italicised in ordinary English text, as in '*et al.*' and in plant and animal Latin names such as *Rattus rattus* (though names of taxa above the level of genus, such as Poaceae, Mammalia, are not italicised), and gene names such as *lacZ* (though the RNA or protein corresponding to the gene is not italicised). Words or abbreviations that have been completely assimilated into English (for example 'e.g.', 'i.e.') do not need italics.

Here are the expansion and translation of some common abbreviations of Latin words.

cf. = *'conferre'* meaning 'compare with'; note no '.' after the 'c' as cf. represents one word.

et al. = *'et alia'* meaning 'and others' (*'alia'* is neuter plural).

etc. = *'et cetera'* meaning 'and the rest'.

e.g. = *'exempli gratia'* meaning 'for the sake of example' (I was once told it meant 'free sample').

ibid. = *ibidem* meaning 'in the same place'; used in footnotes to avoid repeating a citation or reference in the preceding footnote.

i.e. = *id est* 'that is' indicating that what follows is a complete list (not just an example), or another way of putting what has just been presented; often confused with 'e.g.' *q.v.*

q.v. = *quod vide* meaning 'which see'.

sic = 'thus' or 'so' when noting error in a quotation.

viz. = *videlicet* meaning 'namely'; similar to i.e., and now archaic.

In general try to use abbreviations as little as possible, but a few technical ones may be unavoidable. Some of these will be standard for the subject (ATP in biochemistry, DNA almost everywhere) but you must explain any that are not. If you want to use an unusual abbreviation only once or twice and the full phrase is fairly short then spell it out instead of abbreviating: your reader will have forgotten what the abbreviation means by the time you use it a second time. If you need an abbreviation frequently throughout the text, or a few times close together, then define it where it first appears in the text: 'standardised Needham coefficient (SNC) ...'. This sort of parenthetic abbreviation is a warning to the reader 'learn this now as I am going to use it again'. You can convey this message more strongly by putting the abbreviation before the full version: '... HTF (heat transfer factor) ...'. If you use several abbreviations sporadically it may be better to have a section for them after the Summary and before the Introduction.

Dates and clock time

In history and archaeology dates are referred to a base, commonly the BC/AD divide (Before Christ / Anno Domini), or the equivalent

BCE/CE divide (Before or in the Common, Christian, or Current Era). There is no year 0; 1 BCE is immediately followed by 1 CE. The abbreviation BP (before 'present') is referred to 1950 AD (or CE).

Be aware of the ambiguity of a date in the form 9/11/2001. In the UK this would mean 9 November, but in the USA it would be 11 September. Perhaps for this reason many journals require dates in the form '21 January 1995' (day month-as-text year). The month is often abbreviated, for example 2 Apr 2003, or 2-Ap-03.

Clock times can be confusing too: 11:59 p.m. and 23:59 or 2359 (some journals require the colon to separate hours and minutes, and minutes and seconds; others prohibit it) are 1 minute before midnight in the 12-hour and 24-hour systems, but are often followed by 12:00 p.m. and 2400. As the new day begins at this point one might have expected midnight to be 0:00 a.m. and 00:00 or 0000 (24-hour clock). You could use 24:00 to signify the end of a day and 00:00, the identical clock time, to signify the start of a day: 'The experiment took one day, beginning at 00:00 and ending at 24:00.' There are extra opportunities for misunderstanding at midday with the a.m./p.m. system. Most journals expect the 24-hour clock.

A common format for combined date and time is YYYY-MM-DD T hh:mm:ss.s TZD, for example as in 2009-05-16T18:47:30.45+01:00, where the 'T' separates the date from the time, and '+01:00' shows the time zone difference (TZD) from UTC (for Coordinated Universal Time, a compromise between the English acronym CUT and the French acronym TUC for Temps Universel Coordonné). Atomic behaviour determines UTC which is thus invariant. Greenwich Mean Time (GMT) is based on the hypothetical average day, the actual day length varying slightly with the behaviour of sun, Earth, and moon.

Typefaces and fonts

Typeface design and nomenclature are specialised subjects with five centuries of history. Some of the old technical terms are still in use, and you will come across them, so here is a brief account of the commonest ones. 'Typeface' refers to the artistic design; 'font' or 'fount' to the implementation in metal or computer code (similar to the difference between a patent protecting an idea, and copyright protecting an implementation). Both 'font' and 'fount' derive from the Latin 'fundere' and thence French 'fondre' meaning to melt, which also gave us 'foundry' (referring to the pouring of the hot molten alloy of lead and bismuth used to form the individual pieces of type) and the Swiss 'fondu'.

Since the early days of movable (reusable) type the typesetter (compositor) used a 'case' – really two trays like a small shallow open suitcase. Base and lid were divided into rectangular compartments each of which held the loose type of a single character. Compartments in the horizontal (and more accessible) 'lower case' held the small letters and other frequently used characters. Compartments in the sloping, and less accessible, 'upper case' lid contained capital letters and other less frequently needed characters such as punctuation. ▾ A complete case housed a 'font' of a specific variant of a typeface and size. Variants of the basic typeface giving emphases of several kinds might include *italic*, **bold**, small capitals, and combinations of these, perhaps with underlines. The size of characters was specified by the height in 'points' from the base of the descenders of letters such as 'p' and 'g' to the top of the tallest ascender of letters such as 'b', 'd', and 'l'. The size of the printer's movable type 'point' has varied a bit around 72 per inch. For computers, the PostScript (USA) point is now defined as 0.351 mm, in Europe the point is 0.376 mm (Garfield 2010). This text is 10 pt. For each size and variant of type a separate 'case' was necessary. Together the variants in style and size to a common design formed a 'typeface family'.

▾ The arrangement of compartments was standardised soon after the spread of printing with moveable type so that typesetters – members of what was at that time perhaps the most skilled and best-paid manual craft – could move between printers without having to learn new layouts. The compartments differed in area: 'e' being the commonest letter in English had

PUNCTUATION AND TYPOGRAPHIC DETAILS

the largest compartment in English typesetters' lower cases. The order of letters was a blend of frequency of use and alphabetic order. The arrangement of upper case characters was mainly alphabetic. The original upper case layout lacked 'J' and 'U' which, before pronunciation changes made the new letters necessary, were represented by 'I' and 'V'. To avoid altering the layout of all the letters after 'I' the new letters were put after all the others (at the end of the alphabet) in the bottom right of the case.

The typesetter picked characters from the cases without looking at them, just as a professional typist does not look at the keyboard, and oriented the type by the feel of a groove on one side of the base. When he checked a completed page visually he saw mirrored characters and had to learn to read them thus. The expression "mind your 'p's and 'q's" may have been an admonition to apprentice typesetters who were still unsure of this skill (they might have been reminded of 'b's and 'd's, and of 'u's and 'n's too). In the early days the typesetter packed the individual pieces of type and spacing 'furniture' into a frame which was then clamped and used for printing. Later, for commonly needed phrases, a 'stereotype plate' with characters projecting above it was produced. This reusable plate was a cliché (many printing terms are French). The term cliché has been borrowed for any stock phrase used as a unit.

Computers revolutionised printing processes and now dominate them. The craft of typesetter has almost disappeared as authors have taken, or been driven, to the word-processor, but we still use 'uppercase' (now commonly as one word) and 'lowercase' letters.

Computer type characters are stored in two ways: 'bitmapped' and 'vectored'. A bit-mapped character is stored as a matrix of points that are to be either foreground or background colour (commonly black on white), while a vectored character is stored as a set of instructions to draw lines from here to there. 'TrueType' and 'Postscript' (= 'Type 1' = 'Adobe') formats are among the commonest for vectored fonts. Bit-mapped characters need separate storage for every size, and this becomes wasteful at large sizes. A specific size, such as 29 pt, may not exist at all. A vectored character can be printed at almost any size you choose, and at large sizes it looks better than the corresponding bit-mapped one. Vectored fonts are much the commonest.

There is the astonishing total of over 100 000 typefaces in The Fontshop Directory (Garfield 2010), and several computer programs that allow you to modify an existing design or to produce a wholly new one. Some typefaces, such as the cursive scripts and the simply weird, you can ignore for scientific – and perhaps for all – purposes. Special symbolic and mathematical typefaces may be necessary: it depends on the sort of work you do. Classification of typefaces is worse than biological classification: not only are there many exceptions, but there is only a weak underlying inheritance mechanism to push individuals into groups. The British Standard uses the long-established (Maximillian) 'Vox' system, while commercial suppliers have developed their own systems.

Three common categories are these:

- 'Serif'. Serifs are the sideways projections from the end of a stroke, like the pediments on a column. Times New Roman, **Bodoni**, Garamond, and Sabon families are examples. This book uses 10-point Rotis Serif for general text.
- 'Sanserif'. French 'sans' = 'without'. Sanserif is often abbreviated to 'sans'. The **Helvetica, Univers, Frutiger, Arial** (a widely available tweaked clone of **Helvetica**), and Calibri families are near ubiquitous. All but Calibri use a square for the dot above 'i'; **Helvetica** and **Univers** have a horizontal cutoff to the limbs of 'c', 'e', and 's'.
- 'Monospace' or 'typewriter'. Courier New is an example.

Most common typeface families have at least four variants: 'roman' or 'regular'; *'italic'* or *'oblique'* or *'sloped'* or *'slanted'*; **'bold'**; and ***'bold italic'*** (or italic equivalents).

Typefaces with serifs are generally held to be easier to read in running text than sanserif ones. Times New Roman is as good a choice as any for general scientific work. A complete set of roman alphabet, numerals, and punctuation marks in Times New Roman occupies less horizontal space than most other typefaces (Stanley Morison, with Victor Lardent, designed it for *The Times* newspaper, which had narrow columns) and it is more robust than most other typefaces (important when type was cast in metal, though less so nowadays) and is easy to read. Modern computer implementations of Times New Roman contain numerous special characters designed in the same style as Morison's original. The publisher of your article will probably choose a text typeface for the whole journal, to give it a uniform appearance, so it may matter little what you have used in your text. In talks, posters, and symposium and conference reports that are to be produced direct from your electronic file, you may have a choice.

Sanserif typefaces are often advised (and modern ones were designed for) clear display of isolated phrases or short sentences, such as the labels on graphs, warnings, direction signs, and similar matters. **Helvetica** and its derivatives are common for commercial work; Calibri was adopted by Microsoft for all purposes (Garfield 2010), and may therefore be the typeface with greatest exposure to readers. Some journals specify the typeface to be used in tables and figures, where sanserif fonts are most likely to be useful, but many do not, so your choice may survive the editorial process.

Tables with only a few short entries may look best in a sanserif font.

The majority of text typefaces are 'proportional': they allow each character a suitable specific width. Most word-processors adjust these widths to suit particular combinations of characters (compare 'VAV' in which the start and end of the 'A' overlaps the end and start of the enclosing 'V's – a process called 'kerning' – with the same letters arranged as 'AVV' so there is no kerning between the 'V' and 'V' and the three-letter combination then takes more space). In 'monospace' typefaces, however, the characters are all the same unvarying width, simulating a mechanical typewriter which requires the carriage to advance by the same amount after each key is struck. Monospace typefaces may be useful, rarely, in tables when you want to ensure that digits in different rows are aligned. But they are conspicuously different from the other kinds of typeface, and word-processors often allow one to align on a chosen character (such as '.') in numerical entries in a table. There are few monospace typefaces and in print they occupy much more space than do proportional ones of the same point size. Avoid them if possible.

The typefaces named so far are all sold as commercial fonts, some are supplied 'free' with an operating system or office suite. Equivalents of some of them may be downloaded without cost: the 'Liberation' and 'Free' sets of typefaces for example. A recent addition to the free fonts is the Gentium typeface family. It takes less space than any of the other fonts listed here.

In Table 5.1 are examples of serif, sanserif, monospace and cursive typefaces, all at 20-pt size so you can see details of the differences in design. The lowercase letter 'ell', the digit 'wun', and the uppercase letter 'eye' (l1I) are grouped between the lowercase and uppercase blocks because these three characters are not sufficiently different in many fonts. In particular, in Times New Roman the lowercase letter 'ell' and digit 'wun' are not easy to distinguish. Gentium is better. Among the sanserif fonts here the lowercase 'ell' and uppercase 'eye' are confusingly similar, and the often-recommended Gill Sans is particularly poor for 'ell', 'wun', and 'eye'.

One serif and one sanserif font, and perhaps a mathematical font, are probably all that you need.

Special characters

Printers have numerous special characters; many of them are rare. A few of the non-alphanumeric symbols on a keyboard, and some of their common names, are:

- '^' caret, hat
- '~' tilde
- '#' octothorp, hash, grid, 'pound'. The '#' is sometimes used to indicate that a following integer is an identifier, as in '#10' meaning 'item identity number 10' that may, for example, be the third (not the tenth) item in a list

153 PUNCTUATION AND TYPOGRAPHIC DETAILS

TABLE 5.1 Common serif, sanserif, monospace (typewriter), and cursive fonts at 20-pt size. Note the differing space needed for the same example text, and the ease or difficulty of distinguishing the characters 'ell', 'wun', and capital 'eye' – l1I – next to one another in the example text (Gill Sans is particularly poor)

Font name	Example
Serif	
Times NR	acdegpqsil1IWAO029&+-*=,.:?
Gentium	acdegpqsil1IWAO029&+-*=,.:?
Garamond	acdegpqsil1IWAO029&+-*=,.:?
Sans Serif	
Helvetica	acdegpqsil1IWAO029&+-*=,.:?
Futura	acdegpqsil1IWAO029&+-*=,.:?
Gill Sans	acdegpqsil1IWAO029&+-*=,.:?
Monospace	
Courier	acdegpqsil1IWAO029&+-*=,:?
Cursive	
Zapf Chancery	acdegpsil1IWAO029&+-*=,:?

'/' oblique, virgule, (forward) slash, (solidus '/'). (The solidus '/', with its origin in money and mathematics, has a shallower slant than the virgule '/', originally marking a break in a line of poetry. Here they are, side by side: '//'. On a keyboard, the 'forward slash' key usually produces a virgule, and the solidus, in the strict sense, is little used any more even as a division symbol. But in many contexts 'slash', 'forward slash', 'virgule', and 'solidus' are used as synonyms

'\' backslash

'!' exclamation mark, or (in mathematics and computing) factorial, shriek, or bang

'*' asterisk, star, or (in computer programming) multiply

'&' ampersand. The '&' symbol is a stylised shorthand Latin *et* ('and') invented by Tiro, scribe to the Roman Marcus Tullius Cicero. One explanation of the origin of the word ampersand is this. By the early 1800s the symbol was taught to children as if it were a 27th letter of the alphabet. When recited (chanted), letters that represented whole words too ('A', 'I', '&') were preceded by '*per se*' meaning 'by itself'. So the alphabet ended with "... X, Y, Z and *per se*, 'and' ". This contracted to '... X, Y, Z, ampersand'. Nowadays some publishers allow it to replace the word 'and' in formal lists such as those of authors in citations and references, but never in text. Items linked by '&' are of equal weight; those linked by 'and' are often of different weights. Typeface designers have treated the ampersand as a playground for creative design. Early versions showed their origin as a stylised 'et': '&', but most modern examples are variations on abstracted as well as stylised '&'. The '&' is sometimes confused with '@' (commercial 'at', 'petit escargot', 'little snail') which used to be used in bills and invoices for 'at a rate of' as in '20 @ £1.99'. Recently it has wandered into other contexts, particularly email addresses, but is still rare in scientific text.

The symbols commonly used to indicate the probability in a test of significance (* for $0.05 \geq P > 0.01$, ** for $0.01 \geq P > 0.001$, *** for

155 PUNCTUATION AND TYPOGRAPHIC DETAILS

TABLE 5.2 Some of the printable characters not usually found on keyboards: Greek letters and capitals (where these do not resemble roman letters), and some mathematical and other symbols. Good word-processors allow you to find these in a matrix or to type the Unicode value and press a designated hotkey. The unique Unicode of the form 'u????' has four hexadecimal digits (letters may be upper or lower case)

Greek character	Unicode	Name (roman equivalent)	Character	Unicode	Name	Character	Unicode	Name
α	03B1	alpha (a)	−	2212	minus,	—	2014	em-dash
β	03B2	beta (b)	±	00B1	plus-minus	‐	2011	hard hyphen
γ, Γ	03B3, 0393	gamma (g)	×	00D7	multiplication	' '	00A0	hard space
δ, Δ	03B4, 0394	delta (d)	√	221A	square root	' '	2009	thin space
ε	03B5	epsilon (e)	Δ	2206	increment	' '	2002	en-space
ζ	03B6	zeta (z)	∂	2202	partial differential			
η	03B7	eta	∫	222B	integral	†	2020	dagger
θ, Θ	03B8, 0398	theta (th)	≠	2260	not equal to	‡	2021	double-end dagger
ι	03B9	iota (i)	≈	2248	almost equal to	¶	00B6	paragraph
κ	03BA	kappa (k)	∝	221D	proportional to	§	00A7	section
λ, Λ	03BB, 039B	lambda (l)	≤	2264	less than or equal to	‖	2551	double bar
μ	03BC	mu (m)	≥	2265	greater or equal to			
ν	03BD	nu (n)	≪	226A	much less than	°	00B0	degree
ξ, Ξ	03BE, 039E	xi (ks)	≫	226B	much greater than			
π, Π	03C0, 03A0	pi (p)				̇	0307	dot over@
ρ	03C1	rho (r)	←	2190	left arrow	̄	0304	bar over@
ς	03C2	sigma* (s)	→	2192	right arrow	̅	0305	long bar over@
σ, Σ	03C3, 03A3	sigma* (s)	↑	2191	up arrow			
τ	03C4	tau (t)	↓	2193	down arrow			
υ	03C5	upsilon (u)	↔	2194	left-right arrow			
φ, Φ	03C6, 03A6	phi (f)	⇌	21CC	right+left arrow			
χ	03C7	chi (ch)						
ψ, Ψ	03C8, 03A8	psi (ps)	∞	221E	infinity			
ω, Ω	03C9, 03A9	omega						

* In Greek text sigma has two forms: 'ς' at the end of a word, 'σ' everywhere else
@ These three 'over' marks are diacritical marks that can be combined with other characters, as in: $\bar{x}, \dot{m}$

$P \leq 0.001$) were introduced by the statistician Yates as cues to footnotes where the ranges were explained, and without an intention that they should become the near-standard that they have. The egregious 'pH' has an enterprising history too. The chemist Sorensen, working for Carlsberg, introduced it in 1909 as 'p$_H$'

indicating the pressure or potential of hydrogen, but the 'H' escaped from its subservient position and came, like the genie from the bottle, to tower over the 'p' which it should have been qualifying as a subscript. It still follows its master at a respectful distance though. In Chapter 7 I argue that the very idea of pH should have been exposed on a hillside at birth.

Table 5.2 lists the Greek characters that may be useful in equations, and some of the commoner other non-keyboard special characters. Each usable character has a unique Unicode: a numerical representation of four hexadecimal (0–9, A–F) digits. Good word-processors allow you to choose most of the commoner special characters from a matrix-like array of hundreds. But there are many characters not in the matrix and these, and those in the matrix, can be got by typing the four-digit Unicode followed by a designated hotkey.

Many special symbols have been adopted for special purposes, particularly in chemistry, genetics, and molecular biology, but the uses are too numerous to detail here. You can find many of them in Style Manual Committee, Council of Science Editors (2006).

6

Frequently misused words and technical terms

The ill and unfit choice of words wonderfully obstructs the understanding.
Francis Bacon, *Novum Organum*, 1620

Style books often contain lists of words that are commonly misspelled or misunderstood. Many of these words form pairs, some with opposite or similar-sounding partners, others with a similar (or identical: 'rowing' and 'rowing') spelling but different meanings. I will not repeat here, except for a few hardy perennials, what is amply and expertly explained in other works. Buy a dictionary of about 1000 A5 pages or so (and use it) and consult Gowers revised Greenbaum & Whitcut (1986) and Fowler revised Gowers (1968).

What follows is in two lists: misused common words, then often misunderstood more technical terms (usually with longer explanations). The technical terms are also listed without details in the first list.

Many well-educated people would consider a lot of the distinctions that follow in the first list to be mere pedant-fodder – they are – but some impede understanding, and these are certainly worth noting. Remember, too, that a characteristic of good scientific style is that it causes no distractions. Why annoy even a few of your readers when you could easily avoid doing so?

Frequently misused words

affect / effect: 'affect' refers to a cause; 'effect' to a consequence. 'Opening the ventilators affects the temperature in the glasshouse' but 'Open the ventilators to effect (cause) a reduction in glasshouse temperature'.

alternative: derives from the Latin for the other of two, but even pedants now mostly admit that it can be used when a choice is to be made among more than two possibilities. But do not confuse 'alternative' (implying a choice to be made once only) with 'alternate' (meaning repeated change from A to B to A to B to A . . .). 'The director was faced with appointing one twin as full-time secretary or, alternatively, alternating them, each part-time, in the job.'

among / between: between two, among more than two. A distinction without much merit (Fowler revised Gowers 1968).

amount: see the technical list.

and/or: an ugly construction avoidable by 'A or B or both'. The 'or' here is inclusive. There is a rarer exclusive 'or', sometimes called 'xor' or 'eor', meaning A or B but not both. Booth (1993) suggests that we need a word 'andor' for the 'and/or' construction. See *either/or* as well.

anaerobic / anoxic: see the technical list.

area: see *region*.

Avogadro: see the technical list.

between: see *among*.

biannual: occurring twice a year.

biennial: lasting two years or recurring every two years

biweekly: is ambiguous so avoid it. Sometimes it means 'twice weekly' and sometimes 'fortnightly'.

brief: see *concise*.

by / from: a frequent source of confusion if used together in the same sentence. 'Streams should be protected from all pollution by legislation'. Having got as far as 'protected from all pollution by' your reader may be expecting the cause ('chemical releases' or something similar) rather than the cure. It may be better to reverse the 'from' and 'by': 'Streams should be protected by legislation from all pollution', though this does put the solution before the problem.

caption / legend: (also *underline* and *cutline*) is the text that titles and gives surrounding information for a table or figure. It is sometimes said that a caption goes above and a legend below its table or figure but there seems to be no truth in this. Some publishers distinguish a caption as a grammatically incomplete sentence, whereas a legend contains at least one complete sentence. Others reserve legend for the key placed on a graph explaining the symbols used in the graph.

case: 'the case that' can often be omitted or replaced by something shorter. 'It is often the case that such a sentence can be made shorter' becomes simply 'such a sentence can often be shortened'.

Celsius: see kelvin in the technical list.

centigrade: see kelvin in the technical list.

certain: as in 'certain samples were rejected' introduces an unsuitable air of mystery: 'some' or 'a few' are better, but best is to specify how many and the cause as in 'three cracked samples ...'.

circa, and its abbreviated versions *ca* and *c.* are best reserved for dates and not as all-purpose substitutes for 'about' or 'approximately'. All three versions of '*circa*' should be italicised in ordinary use.

circumstances: purists point out that 'circum' derives from the same Latin origin as does 'circumference' and means the conditions around. One can then be 'in the circumstances' but not usually 'under' them. Of course words, like children, do drift away from their parents, and the word 'nitpicking' springs to mind. But why cause annoyance? You save three letters too.

concentration: see the technical list.

concise / brief: 'concise' means without unnecessary words; 'brief' means 'short'. A concise account is not necessarily a brief one, and a brief account is not necessarily concise.

correlated / associated: 'correlated' has such a specific statistical meaning that it may be best to avoid it in its general sense in scientific writing and to use 'associated' instead.

citations / references: see the technical list.

colour / colorimeter: 'colour' contains a 'u' that is absent in 'colorimeter' (but in the USA, Noah Webster's spelling reform ensures that neither word contains 'u', and nor do 'rumor' or 'honor' or 'humor' or 'vigor' or most other polysyllabic British English words ending in -our).

comparable: see *similar*.

compare to / compare with: 'compare to' is to liken one thing to another; 'compare with' is to note the resemblances and differences between them. 'Shall I compare thee to a summer's day?' but 'Compared with a summer's day, a winter's day is short and cold.'

compose / comprise: 'the whole is composed of the elements A, B and C' while 'the whole comprises the elements A, B and C'. You can usually replace 'comprise' by 'consists of'.

constant: see the technical list.

continual / continuous: One hopes that the supply of electricity will be continuous (without interruption) and will not suffer continual (repeated) breaks.

control: see the technical list.

dependant / dependent: these are noun and adjective. A lamb is a dependant of its mother ewe: it is dependent on her.

desiccate / dissect: 'desiccate' has a single 's' and two 'c's; 'dissect' has the reverse.

decimate: 'decimate' means to reduce *by* one-tenth. In a mutinous Roman legion one man in ten would be chosen by lot for punishment – a severe but regulated response. The word is so widely misused or misunderstood nowadays to mean reduce *to* one tenth, and in circumstances where it is impossible to tell whether the writer means 'by' or, perhaps hyperbolically, 'to' one-tenth, that it is probably necessary to abandon it. Thus is a good and useful member of the language ignorantly and needlessly slain. I feel about this as I did about the sight of a colleague using a wood chisel to break out a cracked glass tap from a glass flask. All three items were ruined.

differ: see *vary*

dilute: see the technical list.

dissect: see *desiccate*.

disinterested / uninterested: 'disinterested' means impartial, unbiased, not influenced by one's own advantage; 'uninterested' means indifferent, lacking any interest (next step 'bored'). One hopes that a Judge will be simultaneously interested and disinterested. A one-time UK Chancellor of the Exchequer (Finance Minister) declared (April 2002) on the radio that 'it is more than a pity that so many people are disinterested in politics'. One would be happier if there were signs that this were so. Even worse, a former UK Secretary of State for Education made (Feb. 2003) the same mistake. Pinker (1994) records that at one time these two words did mean the same thing, and that the useful difference between them is a recent development. Welcome, then, and protect a new member of the language.

dissolved: see *soluble* in the technical list.

diurnal: from the Latin '*diurnus*' = 'daily' (as are French 'jour', and 'journey', 'journeyman', 'journal') has two meanings: (1) of or during the hours of daylight, the complement of nocturnal: 'mice are nocturnal, but some of their predators are diurnal and prey on them only at dawn and dusk'; (2) daily, of each solar day as in 'diurnal pattern of light and dark'. 'Diel', for daily, avoids this problem.

due to / owing to: 'due to' is an adjectival phrase and must have a noun to agree with. In 'the failure of the experiment was due to unusually bad weather' the failure was due to bad weather. But in 'there is no record of the experiment in that year due to the bad weather', 'due to the bad weather' is trying to attach itself to 'year'. Fortunately 'owing to' is safe in such situations, and 'the result of', 'caused by', and 'because of' wait in readiness. Putting cause before effect will usually help too.

effect: see *affect*.

efficiency: see the technical list.

either/or: an ugly construction avoidable by 'either A or B or both'. Perhaps we need a word 'eithor'? See *and/or* as well.

elevate: this three-syllable word from the Latin for 'raise', 'lift up' has recently escaped from harness and is rampaging about usurping 'raise', 'increase', 'more', 'higher', and general up-ness. It should be recaptured.

emission: see *flux* in the technical list.

ephemeral variable: see the technical list.

expect / anticipate: 'anticipate' means not just expect but to take action in that expectation. The late A. P. Herbert wrote in a more restrictive time that to say that 'John and Jane anticipated marriage' was not the same as saying that they 'expected to be married'.

first / firstly: there was once a 'rule'– I was taught it at school – that required a list to be itemised first, secondly, thirdly, and so on. There seems to be no good reason for this, and nowadays either first, second, third, or firstly, secondly, thirdly are common. But avoid arbitrary mixtures of the two, even that dictated by the old 'rule'.

fewer / less / smaller: 'fewer' is for discrete items, 'less' and 'smaller' for something considered as a continuous quantity. 'Fewer grains of sugar' but 'less sugar'; 'fewer people' but 'a smaller population'. It is probable, though not certain, that the police spokesman who reported that 'our new stop-and-search policy will mean stopping less innocent people' intended 'fewer'. There are some difficulties: 'fewer than three out of every hundred passengers entered the station smoking' but 'less than 3% of passengers entered ...'. In the first example the writer is imagining individual people, just as they were when the data were collected; in the second the result is viewed as a continuous variable. But should it be 'The journey will take fewer than ten days (camping each night)' or 'The journey (as a whole) will take less than ten days ...'?

first order, second order: see the technical list.

flux: see the technical list

former / latter: avoid. The construction requires that your readers have in mind a firmly ordered list of the recent points you have made or facts you have stated. This is probably not true, and they have to go back and reread in order to make sense of former or latter. 'Respectively' at the end of a sentence often has a similar effect: 'The reduction in mortality in adult males and females in socioeconomic groups C and D in Scotland was 5% and 7% respectively'. Is that gender or socioeconomic group?

frequent / regular: 'regular' is needed if the intervals between events in time or objects in space are the same or nearly the same: 'the fence posts were as regular as a film-star's teeth'. But 'frequent' is needed if the interval between events is, in the context, short but not necessarily regular: 'Young birds require frequent feeding.' 'Frequent' is usually restricted to events in time: 'the farm had frequent ditches' probably implies travelling across it and finding closely spaced ditches. 'Regular' has several other uses too, particularly those implying 'standard' or 'normal' or 'ordinary' or 'medium'.

from: see *by*.

gender / sex: the difference between these two has been vigorously debated and has become more important as the meanings of 'sex' have changed in recent decades. Standard dictionaries, being always behind the times, are little help. 'Gender' is related to 'genre' and 'genus': it is a neutral term for kinds or sorts of objects. In an analysis of a survey 'child', 'teenager', and 'adult' human may be three genders. A specific use of 'gender' is for the noun-categories masculine, feminine, and neuter in some European languages. It has been hijacked to refer specifically to the male and female sexes, particularly of humans, which have mainly biological (genetic and physiological) meanings and causes. On this view a person might change his or her gender, which is a cultural category, but cannot change sex (though they might change the associated external appearances). Some writers treat sex and gender as synonyms. One can hope only that these shifting sandbanks will eventually settle into some quasi-permanent geographic feature.

greater: see *higher*.

greenhouse: describes the hopes of the gardener. The word you often need is 'glasshouse'. But 'greenhouse effect' is too firmly fixed to be dislodged, and the panes in a 'glasshouse' may be plastic. If you have a choice, use glasshouse where it is more accurate.

higher, larger, greater: 'higher' has an element of 'up' in space, 'larger' associates with area and volume, 'greater' with relative position. The contrasting pairs are higher–lower, larger–smaller, greater–lesser. Where the thing to be qualified has a clear association then use the appropriate qualifier: 'higher altitude'; 'a larger shirt', 'greater support'. Where there is no clear association of this kind, some editors take a hard line and always use 'greater'. But all three terms are widely accepted in common use: 'a higher/larger/greater concentration is necessary', 'higher/larger/greater profits may imply more expert leadership'. The same considerations apply to 'lower', 'smaller', and 'lesser'. And the pairs 'more–less' and 'better–worse' stand in reserve.

hopefully: is an adverb as in 'we will travel hopefully' (i.e. in hope), but 'hopefully we will arrive later today' is not usually intended to describe 'arriving hopefully' (rather than in a depressed state). Pinker (1994) justifies this use, but it will still annoy more squeamish readers. Best avoided.

hypothesis: a proposition put forward as a basis for testing or further reasoning. (See *theory* and *law* too.) A 'hypothesis' may arise as a prediction from a 'theory'.

hyper- / hypo-: prefixes meaning above (hyper) and below (hypo) some reference value. Because 'hyper' and 'hypo' look so similar, many (most?) readers faced with, for example, 'hyperthermia' and 'hypothermia' need to hesitate while they work out whether fever or frostbite is the more likely attendant. Avoid these 'hyp-' prefixes if possible. (See *inter / intra* too.)

impact: has an element of a sudden, violent, and probably damaging collision, but is too often misused nowadays where all that is intended is a benign effect. 'The impact of the recent extreme weather on river vegetation ...' is acceptable but 'The impact of superabundant insects on fish growth' is probably not. If you find you have written 'impact' then stop and consider whether or not you could express your meaning more exactly by 'reduce', 'increase', 'affected by', 'effect', 'diminish', 'alter' or some other indication of change. 'Impact Factor', however, is correct because almost all the effects are damaging (Chapter 1).

imply / infer: 'the facts imply ...', but it is you who 'infer from the facts ...'

initiate / **instigate**: 'initiate' is to begin, to start; 'instigate' is to incite, to bring about by persuasion, to provoke – a special sort of initiation. But if you feel inclined to use 'initiate' then ask yourself if 'start' or 'begin' would not be better.

insulate / **isolate**: 'insulate' is used where there is an element of severely restricting a flow of something from one place to another – heat or electricity, for example – but 'isolate' is used in a wider context where two things exist without, or are prevented from, interacting. 'Those who live on isolated islands are insulated from events in the wider world.' 'If the insulation on a cable breaks down you will have to isolate the mains electricity supply.' Some languages have only one word for both meanings.

inter- / **intra-**: prefixes. Inter- means between-, intra- means within-. Because 'inter' and 'intra' look so similar, many (most?) readers faced with 'intercellular' and 'intracellular' need to hesitate while they work out which of the alternatives is intended. 'Between cells' and 'within cells' reduce this problem: the increase in understandability more than pays for the increase in length. (See *hyper* / *hypo* too.)

isolate: see *insulate*.

'**it**': is strongly attracted to the nearest preceding noun or noun phrase, and is a frequent cause of distraction or even of ambiguity. 'The bus was in an accident as it drove north' briefly distracts because at first reading it seems to be suggesting that the accident drove north. 'The bus was in an accident with a lorry as it drove north' is seriously ambiguous: is it the bus or the lorry that was driving north (or even both)? 'He', 'she', and more rarely 'they', present similar risks to 'it'.

jargon: special words or abbreviations, or common words used in a special way, are a useful – an unavoidable – feature of communication between specialists in their common field. There is nothing wrong with jargon used for that purpose. But it is often overused, or used outside its proper field either thoughtlessly or, worse, with intent to impress or confuse.

kelvin: see the technical list.

latter: see *former*.

larger: see *higher*.

law: when a theory (*q.v.*) has been repeatedly tested and not rejected it gradually acquires the aura of a fact and begins to be referred to as a law. The hypothesis that 'flow (of electricity, or of water in a conducting or porous medium) is proportional to potential difference and inversely to resistance' passed through the theory stage and is now known as Ohm's Law (electricity) and Darcy's Law (water).

legend: see *caption*.

least worst (option or choice): a choice may be the least bad, but there can be only one 'worst' choice. This is a fairly new but virulent mistake, which resist lest 'most best' tries to invade.

less: see *fewer*.

lesser: see *higher*.

level: most of the useful meanings derive from that of a horizontal plane. But 'level' is widely misused in scientific and other writing as an all-embracing word to refer to 'the-thing-that-I-am-thinking-about-but-am-too-lazy-to-give-a-specific-name-to' as in 'the level of blood sugar increased' rather than 'the concentration of sugar in the blood increased'. Similar considerations attend 'size' (*q.v.*).

light: see the technical list.

limiting factor: see *nutrient* in the technical list.

located: is usually redundant: 'the experimental site was located 10 km from the nearest road'. Why not 'the experimental site was 10 km . . .'?

loose / lose: 'loose' means unrestrained, unattached; 'to lose' is to be deprived of. Lose an 'o' from 'loose' to give 'lose'.

lower: see *higher*.

mass / mol: see the technical list.

mass / weight: see the technical list.

may / might: These have several uses. The clearest is when 'may' is used to show that an action is permitted (and 'may not' that it is forbidden): 'I may not travel with this ticket before 09:00'. The next uses are to show fine shades of likelihood. 'May' states that something is possible, 'might' adds uncertainty. 'He may go tomorrow' (it is possible); 'he might go tomorrow' (but perhaps on the day after). The same distinction arises in conditions. 'If you ask nicely he may tell you' (it is worth trying); 'if you ask nicely he might tell you' (but do not rely on him doing so). The condition may be more stringent and 'may' will not do: 'If his parachute had opened he might have survived' implies that it (and he) did not, so 'may' is illogical. In these situations 'may' and 'might' can be used in the past, present, and future. 'He may have gone' (possible, but I do not know); 'he might have gone' (but decided not to). In practice, 'may' and 'might' have become almost interchangeable and the distinction should not cause you to worry.

maximum: see *optimum* in the technical list.

metre / meter: 'metre' is the distance; 'meter' is the instrument (but in the USA both are 'meter').

might: see *may*.

mineral nutrient: see the technical list.

mitigate / militate: 'to mitigate' is to ameliorate, to make less bad; 'to militate' (usually against) is to have force or effect against.

mol / equivalent: see the technical list.

more or less: asking yourself 'more or less than what?' is often enough to remind you of the usefulness of 'about' or even of 'approximately'.

mortality: means 'rate of death' not just 'death', just as 'velocity' means rate of movement. So 'the mortality was 0.05 yr^{-1}' is correct but 'the rate of mortality was 0.05 yr^{-1}' is not. 'The rate of mortality is 0.05' is useless as well as wrong, because it lacks a unit of time. Also wrong is '... individuals and populations both suffered mortality'. 'Natality' is a rate too. So is 'productivity' (*q.v.*) and most other variables (*q.v.*) whose name ends with '-ity'.

MS: abbreviation for 'manuscript' which comes from the Latin for 'hand' and 'write'. MS is widely (mis)used nowadays for an author's word-processed offering. 'TS' (for 'typescript') is better, but plain 'script' is simplest.

nutrient / limiting factor: see the technical list.

only: is strongly attracted to the word or phrase immediately following it (or, at the end of a sentence, immediately preceding it). If you move it earlier in the sentence for emphasis it may – probably will – stick to the wrong thing. 'His ideas can only be used with caution ...', not rejected, or promoted, or patented, or any other thing that might be done with an idea. Purists would write 'his ideas can be used [but] only with caution ...'.

Not convinced? Try inserting 'only' before each of the five words, and after the last, in 'I shook his hand yesterday'. You get six sentences each with a different meaning.

optimum / maximum: optimum (plural optima) often implies the 'best' combination of factors with conflicting tendencies, whereas maximum is merely the numerical value(s) of the independent variable(s) that give(s) the biggest observed numerical value of the dependent variable. Optimum also has an element of being the best possible in any circumstances, whereas maximum is merely the best in the limited circumstances of the particular work or experiment. If you find you have written 'optimum' then make sure that you really mean it.

or: see *either*.

oral / verbal: oral refers to speech; verbal means 'in words' – either spoken or written. But 'verbal' is often misused where 'oral' is intended. 'Oral' may also be confused (mostly in speech) with 'aural' which refers to hearing. Oral and aural have the same relationship as do lending and borrowing: 'oral' refers to giving and 'aural' to receiving.

order of magnitude: see the technical list.

paradigm has become a vogue word. It originally meant 'a *typical* example, a pattern followed', but has drifted to mean 'the characteristics that define a scientific field at a particular time'. Many, perhaps most, readers will not know this and if you use 'paradigm' you may obscure rather than clarify your intention.

parallel: see *series*.

parameter: see the technical list.

percent / proportion: see the technical list.

plot: see *region*.

plural: see *singular*

prescribe / proscribe: 'to prescribe' is to recommend, to impose; 'to proscribe' is to banish, reject, forbid. Both originate from the Latin word for 'write', but 'pre-' and 'pro-' give them almost opposite meanings.

programme / program: In the UK one follows a 'programme' of study, while Americans follow a 'program'. But in both places the instructions to a computer form a 'program', and the person who wrote them is a 'programmer'. The difference between the UK and USA reflects the early rapid exploitation in the USA of stored-program computers. ▼

▼ The earliest satisfactory digital computer is probably the 'Colossus', designed, built and commissioned in 1944 largely by T. H. Flowers employed by the British Post Office (Copeland 2006). Its objectives were specified by codebreakers at Bletchley Park to help decode German secret messages from the Lorenz series of machines in a code nick-named 'Tunny', broadcast by wireless in teleprinter code. This was a more advanced code than the now well-known 'Enigma'. Its cracking by Bill Tutte, with no example of the machine's structure, was an extraordinary achievement. Eleven Colossi were built. At the end of the war orders were given that the enterprise was to remain secret indefinitely and the Colossi were to be destroyed. Most were, but one or two survived and continued in use into the 1960s. A modern example has been reconstructed from original reports. The Colossus had about 2500 thermionic valves which were widely thought to be unreliable. But Flowers had shown that it was switching on and off which led to failure: once powered up, the Colossus ran nearly continuously, and was very reliable. In the 1940s valves were used widely for amplification, but Flowers was one of the few who knew about their use as fast switches for binary representation. Colossus was digital and electronic; it read, stored and retrieved data, printed results, and was controlled by a program. The program was fixed but a plugboard and hand switches allowed limited reprogramimg. It was thus a genuine digital electronic computer, as the term is now understood. It was in operation two years before the similar and well-known ENIAC, but because of the decision to keep its existence secret the history of this remarkable computer was released only 30 years after its construction.

proportion: see *percent* in the technical list.

proof: given agreed axioms and rules for manipulating them (logic) a mathematician may construct a proof showing that some inference is a necessary consequence of the axioms. But experimental science can never prove or disprove a proposition ('hypothesis') in the real world, though it may be easy to make the probability or likelihood of its truth (falsity) almost indistinguishable from 1.0 (0.0). This resembles the English 'legal proof' in the criminal courts as 'beyond reasonable doubt'.

An example that is said 'to prove the rule' is not proving it at all: it is testing the rule. And a spirit that is '100% proof' was at one time defined by the practical test of being just able to sustain combustion when mixed with an equal volume of gunpowder. (There are several more modern ways of defining it.)

quotient: see *ratio* in the technical list.

range: see *vary*.

rate: see the technical list.

ratio / quotient: see the technical list.

region / area / site / plot: all indicate a geographic place and form a hierarchy. Stick to the same word for the same idea throughout what you write in a single article. It is surprising how often authors wander about among these terms sowing confusion and reaping suspicion as they go.

regular: see *frequent*.

relation / relationship: one may consider 'the relation between mother and daughter', and 'the mother–daughter relationship', but avoid 'the relationship between mother and daughter'.

repeat: means to do again. So a measurement 'repeated three times' has actually been made four times. Perhaps 'we made the measurement four times' will do?

respectively: see *former / latter*.

series / tandem / parallel: items arranged one after the other are in 'series' (consecutive if in time). In contrast are items 'in parallel' (concurrent or simultaneous in time). Two items in series can also be described as '(in) tandem' from the Latin 'tandem' meaning 'at length'. 'Tandem' is increasingly misused to mean 'in parallel': a solecism – a mistake – the spread of which should be vigorously opposed as it means the opposite of what was intended.

sex: see *gender*.

shall / will, and should / would: are used to deal with conditions where there is a simple wish or unforced resolve, and with those where there is an obligation or command. We are in deep water here. Usage differs for the first person (I, we) and other persons ('thou', he, she, it, you, they). Let 'I' and 'you' represent these groups. In southern England, at present:

Shall / will		Should/would
I shall go; you will go.	(unforced resolve)	I would go, you would go.
I will go; you shall go.	(obligation, command)	I should go; you should go.

This is over-simple, however, and is inaccurate in other regions. Fowler & Fowler (1906) in 'The King's English' need 23 pages to consider this topic: "the idiomatic use [which] ... comes by nature to southern Englishmen is so complicated that those who are not to the manor born can hardly acquire it".

The basic state above is certainly incomplete. To it should (obligation) be added numerous exceptions and special cases, and different uses in different regions. It is said that a Scotsman who got cramp while swimming in a lake in southern England shouted 'I will drown and nobody shall save me'. The polite English onlookers, respectful of the rights of the individual, left him to do so.

For a language as widely used as English one may doubt the wisdom of trying to promote any construction as complex as this. Evolution is simplifying usage. Meanwhile, do not worry about this quagmire.

should / would: see *shall / will*.

significant: has developed such a specific meaning in statistical analyses that it is best avoided in its general sense. Replace 'This is a significant finding' with 'This result has clear [or important or notable] consequences' or something similar.

similar / comparable: one thing can always be compared with another (though it may make little sense to make the comparison). If, having done so, you find many features or values the same, or nearly the same, then the two things are similar.

singular / plural: 'Number' is a collective noun and the cause of confusion. When it is part of a composite subject then it needs a plural verb as in 'a large number of seeds were planted' (the composite subject is underlined). But when number is a simple subject then the verb should be singular: 'the number of seeds planted was large'. You can avoid the problem by writing 'we planted many seeds'. But the 'correct' 'There has been a number of turning points ...' will read awkwardly to many readers who will not be distracted by the 'incorrect' 'There have been a number of turning points ...'. Again, you can avoid the problem with 'There have been several turning points ...'.

The values recorded in a table are individual though the table can contain many such values: 'Values are the concentration of ...' rather than '... are concentrations of ...'.

English spoken numerals are inconsistent: 0 °C is 'zero or nought degrees Celsius' (zero seems to be plural), 0.5 cm is 'zero point five centimetres' or just 'point five centimetres' (plural again) but may also be 'half a centimetre' (singular), 1 cm is 'one centimetre' (as expected), 1.0 cm is 'one point zero centimetres' (plural yet again), and 2 cm is 'two centimetres' (plural as one would expect). German has the same difficulties, and so has Swedish (H. Rydin, personal communication, September 2002).

The underlined word of the following singular and plural pairs is frequently misused:

agenda	agenda; ('things to be done') singular 'agendum' is now pedantic
alga	algae
bacterium	bacteria
criterion	criteria
datum	data (from the Latin word '*dare*' for 'to give'. You may wonder whether or not the hard-won results of an experiment should be described as 'given'. And whether a 'donation' is not just a slightly pretentious gift.)
die	dice (the gaming device)
flagellum	flagella
formula	formulae (losing ground to formulas)
maximum	maxima (not yet maximums)
medium	media (a lost battle in the worlds of newspapers, radio, and TV, but that is no reason why bacteriologists, printers, and artists should surrender)
optimum	optima (not yet optimums)
phenomenon	phenomena
stoma	stomata (not stomates; from the Greek for 'hole')

Many, but not all, words ending in '-ix' and '-ex' have two plural forms. If the cartoon character Asterix had had a brother then together they would probably have been 'the Asterices' in keeping with the Roman times in which they lived. But two washing machines are 'Bendixes'. Examples:

appendix	appendices	appendixes
index	indices	indexes (lists of items in alphabetic order are 'indexes'; superscripts and summarising values are 'indices')
prefix	prefixes	
vortex	vortices	vortexes

site: see *region.*

size: as in 'proportional to its size ...' is another of those dangerously vague words – such as 'level' (*q.v.*) – that can usually be replaced by a more specific one. For example 'length', 'area', 'volume', 'dry mass', 'population'. 'The size of the sample was doubled' is correct if it went from 10 to 20 items, but would be unambiguous as 'the number in the sample doubled ...'. But if its length were doubled from 10 to 20 cm then it should be 'The length of ...'. And if the phrase refers to a square sample, perhaps of vegetation, of side 1 m, then do you mean the side was doubled to 2 m (and area to 4 m^2) or do you mean the area went from 1 to 2 m^2 (side from 100 to 141 cm)? A related common difficulty is seen in 'the surface was 4 m square' which ought to mean that its area was 16 m^2 but may be intended to show that the side was 2 m: '4 square m', and '4 m^2' (note that the order of unit and index are reversed), mean the same thing and are not the same as '4 m square'.

smaller: see *fewer* and *higher.*

soluble: means capable of dissolving. If a substance is already in solution then it is 'dissolved'.

split infinitive: by the time Star Trek's mission statement was 'To boldly go ...' the objection to the split infinitive was already derided as out-of-date or baseless. Pinker (1994) records that the 'rule' was invented to try to emulate the precision of Latin in which the infinitive is a single word that cannot therefore be split. But a few readers may be irritated if you overuse the licence thus available.

straight / strait: 'straight' is undeviating; 'strait' is narrow, as in 'the Straits of Dover' and 'straitjacket'. The 'strait and narrow way' is tautologous ('strait' and 'narrow' mean the same thing). 'Straight is the path and strait the gate' is the shortest road to a narrow entrance.

strain: see the technical list.

stress: see *strain* in the technical list.

tandem: see *series*.

that / which: the details are intricate. Fowler & Fowler (1906) in *The King's English* take 27 pages to discuss the details. But the main line is clear. Grammarians distinguish a defining clause (without which a sentence would not make sense) from a commenting clause (which can be omitted without destroying the sense). A defining clause is started by 'that' (or, outside the USA, by 'which' or 'who'). A commenting clause should be lead by 'which' (or 'who') and is usually introduced (and, if not ending a sentence) ended by a comma. Compare the three following sentences.

'The notes that [or which] were taken from the file are still missing.' Here 'that [or which] were taken' is essential because it identifies the notes, so the clause is a defining one.

'The notes must be replaced in the file, which is still accessible.' The clause containing 'which' is commenting.

'The notes must be replaced in the file that is still accessible.' Here 'that' must be defining and therefore implies that there are other potentially suitable but inaccessible files.

theory: a proposed explanation of one or more facts or phenomena assumed to be true. (See *hypothesis, law,* and *proof* too.)

TS: see *MS*.

up, up with (and similar constructions): are redundant in phrases such as 'free up', 'meet up', 'meet with', and the doubly irritating 'meet up with'. Do not add a preposition to words that already imply one.

variable / variate: see the technical list.

vary / differ / range: in its statistical use 'vary' draws attention to the fact that instances of a variable tend to differ in numerical value, and 'variance' is one measure of this. In common use, several numerical values may 'vary' but two can only 'differ'. Beyond that there is often an element of time in 'vary' that it may be useful to encourage: 'The pH varied during the year' but 'The pH differed among the 20 sites'. If specific numerical values are given then 'range' can help: 'The pH ranged from 3.6 to 5.8' (during the year or among the 20 sites). Use the word 'to', not the symbol '–' to link the extremes, to avoid confusion with negative values.

verbal: see *oral*.

viscous / vicious: 'viscous' describes a fluid that, as a result of internal friction, resists movement (often conspicuously); 'vicious' is the adjective derived from 'vice'. The recent increase in frequency of this mistake may be no more than dyslexia in action.

will: see *shall / will*.

would: see *shall / will*.

which: see *that*.

while: is best reserved for situations in which the essence is that an event is contemporary with another, as in the carol 'While shepherds watched their flocks by night . . . the angel of the Lord came down'. When two events are contemporary but not otherwise connected you can use 'and' or 'also' or similar links. 'Shepherds watched their flocks by night . . . and the angel of the Lord came down' shows different actions (the shepherds watching and the angel descending) but not necessarily linked ones. You may think that this version lacks something in poetic virtue when compared with the original, as well as having a different meaning.

yes: a useful alternative to 'absolutely' when assenting to a question.

QUESTIONS:
- Do you know the difference between the following frequently confused pairs of words?
- Complement / compliment; council / counsel; draft / draught; discreet / discrete; lead / led; lay / lie; passed / past; practice / practise; principal / principle; stationary / stationery?

Frequently misunderstood technical terms

abscissa / ordinate: in the graphical context, the abscissa is the *horizontal* distance from a point to the vertical axis of a two-dimension graph. 'Abscissa' is also used for the corresponding co-ordinate, though if the vertical axis is plotted at some numerical value other than zero on the horizontal axis these definitions conflict. By extension 'abscissa' is also widely used for the horizontal axis itself (and I use it in that sense in this book). The abscissa is also often inaccurately called the x-axis, though there is no necessity to use the symbol 'x' for the variable allotted to the abscissa. If the relationship between the variable on one axis is assumed to depend on the variable on the other axis, then the independent ('causal') variable should usually be allotted to the abscissa (unless the independent variable has a natural vertical meaning, such as height, depth, altitude).

Similar caveats apply to 'ordinate'. The ordinate is the *vertical* distance from a point to the horizontal axis of a two-dimension graph, and for the coordinate value itself. 'Ordinate' is also widely used for the vertical axis itself. It is often inaccurately called the y-axis. If dependence is assumed, then the dependent variable should usually be allotted to the ordinate axis.

amount / concentration: amount is an extensive property while concentration (amount volume^{-1}) is an intensive one. This is an important difference. See *concentration* in this list too.

aerobic / oxic: see the next entry.

anaerobic / anoxic: 'anaerobic' describes metabolism conducted without free molecular oxygen while 'anoxic' describes an environment without free molecular oxygen as in 'methane is produced anaerobically and only in anoxic conditions'. The opposites 'aerobic' and 'oxic' are similarly differentiated.

Avogadro constant / number: Avogadro's constant N_A = 6.02314179e23 mol^{-1}. Its units are mol^{-1} (the number of entities – atoms or other particles – per mole). It is therefore a constant with units and not a dimensionless number. Avogadro's number is the dimensionless numerical value 6.02314179e23. Note that the name is Avogadro not Avogardro.

concentration: usually an amount (mass or mole) per unit volume i.e. an intensive quantity. It is a pity that 'concentration' has so many syllables: if it had only one then the temptation to (mis)use 'level' would be smaller. In graph axis titles and table column or row titles, where space is limited, consider using the bracket convention: [Ca^{2+}] for 'Ca^{2+} concentration'. See *amount* in this list too.

citations / references: support from another work is shown in the text by a cue to a reference where full bibliographic details can be found. Together the cue and reference form a citation, but in common practice the cue alone is often called the citation. In scientific articles citations are usually 'author–date' or 'numbered', and the fuller details are collected in a References section at the end of the article. See Chapter 11 for details.

constant: depends on the context in which it is used (as does *control, q.v.*). There are universal mathematical constants such as 'π' and 'e'; there are basic physical constants, that we usually assume are the same everywhere and at all times, such as the speed of light in a vacuum c, the gravitational constant G, the gas constant R, Planck's constant H. Some physical and chemical coefficients are constant in defined conditions but change as the conditions change. Examples are the acceleration due to gravity (weight per unit mass) g, the density of water (it depends on temperature). Finally a 'constant' may be chosen arbitrarily as a reference point for a particular experiment only.

control: implies a base, often an untreated sample, against which others can be assessed. But often or usually 'control' is too vague or uninformative, except (like 'weed') of the user's intention, and is best avoided. There are so many possible sorts of 'control': for example experimental, analytical, lacking part of a treatment, receiving some reference treatment. You can usually be more specific: for example 'unclipped', 'untreated', 'reagent blank', '10 °C treatment'.

In English 'to control' (the verb) means 'to constrain within desired bounds'. In some other languages the equivalent of 'to control' can also mean to inspect, examine, measure. It does not have this more general meaning in scientific English.

dilute: a solution, S, the concentration of which begins as 1.0 mol l^{-1} and is described as being 'diluted 1 : 4' is actually diluted 5 times to 0.2 mol l^{-1}. In any case, diluting 1 : 4 (adding 4 volumes of water to 1 volume of S) does not give the identical concentration got by diluting 1 volume of S with water to 5 volumes, because the final volume is not just the sum of the volumes of S and of water: there is shrinkage or, for compounds such as ethanol, acetone, and ammonium hydroxide, expansion during dilution.

Notice too that 'three times less concentrated' than 1.0 mol l^{-1} is ambiguous: does it mean 0.33 or 0.25 or even (unlikely) -2.0 mol l^{-1}?

efficiency: in the strict sense a quotient (*q.v.*) of two values with the same (or no) physical dimensions (see Chapter 7) so that the efficiency itself is dimensionless. 'Photosynthetic efficiency (energy content of new growth / radiation absorbed) was 0.012.' But efficiency is often used more loosely for a quotient that has dimensions as in 'The winner of the petrol consumption tests averaged 127 km l^{-1}: a surprisingly high efficiency.' It should have been '... implying surprisingly high efficiency'.

ephemeral variable: a variable, the present state of which cannot be preserved for later measurement. Temperature and radiant flux density are examples; pH and redox potential should usually be treated as ephemeral.

first order, second order: main and subsidiary effects. Having effects that are at least an order of magnitude (*q.v.*, factor of 10) different in size.

flux: a rate of flow. The rate of flow across a surface is a 'flux density', but 'flux' alone is often used for this concept. 'Net flux' is the difference between 'influx' and 'efflux', where the context has to reveal what is 'in' and what 'out'. The word 'emission' is often used in place of both efflux and net efflux, and for both amount and rate. There is no common counterpart for influx. Better avoid 'emission'.

Kelvin: temperature scales are defined by two fixed points. The Fahrenheit scale fixed points were, at the lower (colder) end, a mixture of ice, water, and ammonium chloride, and, at the upper (warmer) end, human body temperature. This range was divided into 100 equal steps, hence 'centigrade'. (Any scale divided into 100 equal steps is 'centigrade'. The so-called Centigrade scale is a misnomer for the Celsius scale.) Later scales mostly used as fixed points the freezing point (ice point) and boiling point (steam point) of water. Fahrenheit is a centigrade scale, later found to give numerical values of about 32 and 212 degrees for the ice and steam points.

For operational measurements a physical property is chosen and on a centigrade scale its numerical values at the lower and upper points are 0 and 100 degrees. Between those points 100 equal divisions of the physical change between the fixed points is each a degree. Three physical properties in common use are the length of a mercury capillary above a reservoir, the electrical resistance of platinum, and the potential difference of a copper–constantan thermocouple. There are many other possibilities. These scales are all empirical – they depend on a physical property – and in the middle of the range (at about 50 degrees) at the same temperature they deviate from one another by as much as a degree.

What is needed as a reference is a temperature scale that is minimally dependent on the properties of any substance. It was found that the smallest empirical deviations were found among thermometers based on the pressure of several gases at constant volume, and that as the pressure was reduced (as the temperature went down) the differences between gases decreased. All gases when extrapolated pointed to a common origin for zero pressure at close to −273.16 on a centigrade scale defined by the ice and steam points. This is still an empirical observation. The next step was to *define* this absolute zero point on the perfect gas scale of thermodynamic temperature as exactly −273.16, thus establishing one fixed point. The other fixed point was defined as 0 at the triple point of liquid water, ice, and water vapour (found to be close to but more stable than the ice point). The kelvin (K, no '°') is then 1/273.16 of this scale. A 'K' is thus analogous to a length. When a temperature is said to be, for example, '293 K' it means 'the temperature is 293 kelvins above absolute zero'.

Finally the Celsius scale, named after the eighteenth-century Swedish astronomer Anders Celsius, is now *defined* with its zero at 273.16 K, and with the same size of degree as the kelvin, so that $t / °C = T / (K - 273.16)$, where t is the temperature on the Celsius scale, and T on the perfect gas (thermodynamic) scale.

These definitions are now fixed, but improved measurements have shown that between the triple point and steam point are only 99.7 K (and therefore °C too) and further small revisions in this numerical value are likely. The definition of the K and °C is not and will not be affected. A longer account of these complexities is in Adkins (1987).

It is possible that the unit of temperature may eventually be redefined in terms of the difference in the mean energy of molecules, but we are some way from that.

If differences of a degree or so are not important in your work then uncorrected empirical measurements are adequate; if better accuracy is needed then tabulated or automatic electronic corrections should be applied.

For most bioscientists and most of the time the Celsius scale and °C are suitable, but thermodynamic reasoning will require the K.

light: there are three main ways of considering light that falls on a surface. First, irradiance is the flux density of energy onto a surface (SI units W m^{-2}). Second, photon flux density (PFD) is the flux of photons on a surface (SI units mol m^{-2} s^{-1}). The non-SI unit E, for Einstein, is sometimes used for the mol of photons. The energy carried by a photon is inversely proportional to its wavelength, so you can look on PFD as irradiance to which a weighting inversely proportional to wavelength has been applied. Given the spectral distribution of energy in the light then you can inter-convert irradiance and PFD. If you are interested in the energy balance of an object you will probably use irradiance; if your interest is in photosynthesis or other photochemical processes you may choose PFD. The standard irradiance-measuring device uses the heating effect of light falling on a dull black surface, approximating a perfect radiation absorber. You can get PFD by placing between the source and the measuring device a filter that intercepts more strongly in the blue (where photons carry more energy) than in the red, thus weighting the energy spectrum. Practical devices use semiconductors with filters to give them the correct response to wavelength.

The third concept, illumination, is relevant if your interest is in the ability of humans to see things. It takes the energy in different parts of the visible spectrum and weights it in proportion to the internationally defined spectral sensitivity of the average light-adapted eye of the young adult human (SI units for the illumination flux density: lm m^{-2} or lux). The cheapest and most robust device for measuring light used to be the silicon barrier layer photocell which can be filtered to give the correct weighting for illumination measurements. If you have a light source with unvarying spectral distribution then you may use illumination to measure the relative irradiance or relative PFD, but illumination measurements and their units are otherwise of little use to most bioscientists.

The term 'light intensity' is a property of the light source, and is not often needed by bioscientists because they are usually interested primarily in what has reached an organism, for which one of the concepts irradiance, PFD, or possibly illumination, is needed.

mass / mol: the amount of a substance can be reported either by its mass or as chemical moles (number of particles; 1 mol contains Avogadro's number, 6.02314179e23, of particles). If one wishes to compare different chemical substances – for example amounts or concentrations or rates involving C, CO_2, and CH_4 – then it is sensible to use mol (or a submultiple), because 1.0 mol of CO_2 contains 1.0 mol of C, and so does 1.0 mol of CH_4. This principle is used to report light flux of photons (SI units mol m^{-2} s^{-1}) which is needed to calculate the efficiency (q.v.) with which light is used to fix CO_2 molecules.

mass / weight: mass is a measure of the amount of matter (its 'muchness') and is what determines inertia; weight is a force produced when gravity acts on a mass. Before the space age everyday experience was of things all in approximately the same gravitational field, so a given mass would produce nearly the same weight wherever you were at the same distance from the Earth's centre. The practical method for measuring mass was (and is) to compare the weight (a force) with that of a standard mass in the same gravitational field. This process is correctly called weighing, but is usually used as a surrogate to estimate mass (muchness) rather than weight (force). Spectacular pictures from space capsules and special effects in films have made the distinction between mass and weight part of everyday experience – at second hand – so the difference is now more widely recognised.

The formal relation is $w = mg$, where w is weight (force), m is mass, and g is usually called the acceleration due to gravity. But it follows that $g = w / m$, i.e. that g can be understood as the weight per unit mass at a particular place. The value of g depends on where you are; it is not the same as the universal gravitational constant, G in the relation $F = G\, m_1\, m_2 / x^2$, where F is the gravitational force between the masses m_1 and m_2 a distance x apart.

In 'The weight of the 50 kg orangutan was enough to break the branch ...' weight is used correctly for the force that caused the break, while 'The average mass of an adult orangutan is 50 kg' emphasises the muchness of the animals.

mineral nutrient: nearly always an error when used by bioscientists. In the context of geology and soils, a mineral is a solid with a fairly specific crystal structure and at least a generic chemical composition. Calcite is a mineral, so are hornblende and the numerous clay minerals. In the mineral (solid) state such substances cannot easily get into plants. 'Dissolved nutrient' would usually be closer to what is intended. But see *nutrient* for further reservations and then perhaps prefer 'inorganic solute'.

mol / equivalent: a diminishing problem as the concept of the equivalent has not been taught in most schools for decades. But there is a related difficulty. The molecular weight of Ca is 40, and so, to a close approximation, is that of the cationic form Ca^{2+}. So a solution containing Ca^{2+} at a concentration of 40 mg l^{-1} would contain Ca^{2+} at 1.0 mmol l^{-1}. But if one wished to consider the electric charges in the reaction $Ca^{2+} + 2\ Cl^{-} \rightleftharpoons CaCl_2$ then it may be more useful to consider an 'equivalent' of half Ca, represented as '½ Ca^{2+}', carrying a single positive charge. The same solution as before, considered as ½ Ca^{2+}, has a concentration of 2.0 mmol l^{-1}.

nutrient / limiting factor: terms that are often abused. They should be used only in the range of supply of a substance that has been shown in the particular circumstances to increase the response. If the supply of the substance is increased steadily while holding everything else unchanged then for most substances there will come a point where the response no longer increases, and a further point where the response begins to decrease – the substance has become toxic. If the supply of a substance shows that it is a nutrient then by the same evidence it is a limiting factor. Some factors such as temperature and water can be limiting but are not usually considered to be nutrients. More than one factor can be limiting at the same time. For example, the rate of photosynthesis may respond positively to increase in either light flux or carbon dioxide supply or to both. See Clymo (1995) for details.

order of magnitude: bigger or smaller by a factor of ten. 'Variation as a result of pressure differences was two orders of magnitude smaller than that owing to temperature' meaning that, within the ranges experienced, variation owing to differences in pressure was about 1% of that owing to variation in temperature. 'Order of magnitude' is often used to indicate an approximate value prior to saying that the smaller value will be ignored, and it is this vagueness that usefully sweeps away the need for extensive explanations and qualifications.

ordinate: see *abscissa*.

parameter: is used differently in mathematics, in statistics, and in physics. The widest use (relevant to a majority of bioscientists) of parameter is for a distinguishing or defining feature – a value that is constant in the situation being considered. The difference between a variable and a parameter is most easily seen in an example. Imagine a graph showing the yield of a crop (on the ordinate, dependent, vertical, y-axis) plotted against the density at which you added nitrogen (on the abscissa, independent, horizontal, x-axis). The data seem to be scattered about an imaginary straight line pointing up and to the right. As you want to predict crop yield from a knowledge of the nitrogen added, you fit a linear regression, to the equation $y = bx + a$. The quantity 'x' is a variable (it is known exactly) and 'y' is a variate (with a probability distribution causing the scatter): you can show them as the axes on a graph, x as the abscissa, y as the ordinate. The quantities 'b' and 'a' are parameters: they select the specific line that interests you from the infinite number of possible straight lines in the space defined by the x-axis variable and the y-axis variate. It is customary to use the early letters of the alphabet (a, b, c ...) to represent parameters, late letters (u, v, w, x, y, z) for variables and variates, and mid-alphabet (i, j, l, m, n) for indices (counters and identifiers). If what you are writing about is being considered as on the axis of a graph then it is a variable or variate (*q.v.*), not a parameter.

Like beauty, being a parameter is in the mind of the beholder. Imagine again that you have several lines on your yield graph (for experiments in different climates) and you notice that the lines with steeper slopes also have larger intercepts on the ordinate i.e. at $x = 0$. How useful it would be to be able to predict the slope in a known climate from a single measurement with no nitrogen added i.e. from the value of 'a'. So you plot a new graph showing the values of 'b' on the new ordinate, and the values of 'a' on the new abscissa. You are now considering what were parameters 'b' and 'a' as variate and variable to be plotted on graph axes. Whether an entity is a parameter or a variable or variate depends on how you wish to treat it.

Many people suppose that parameter is synonymous with variable (and with variate). It would be a pity to slaughter by ignorance yet another useful distinction.

percent / proportion: percent uses comparison against a base that has the value 100. The base may not be obvious: '80% were caught at the first attempt and 40% were male'. Is that half those that were caught? Or is it 4/10 of the whole group? Proportion sets the reference group at 1.0 instead of 100. Percent is useful when the values are needed to only one or two significant digits: (32%, 40%) thus avoiding the '.', but if there are fewer than say 25 in the sample then percentages can mislead. For example, 5 out of 13 appears as 38%, but 6 would have been 46%: one extra has added 8%.

If the sample contains fewer than 100 values then the percentage value does not justify a decimal point (54%). For 100 to 10 000 no more than one digit after the decimal point can be justified (54.4%). See Table 8.2. Note that the '%' symbol behaves as a unit (as do °C and K) and ought therefore to be separated from the numerical value by a space. Many journals (and this book because that is the publisher's house style) leave no space though.

rate: is so much per unit time. It is a variable like any other, and can be qualified, for example as 'great' or 'small' but should not be 'fast' or 'slow'. More serious is whether the rate is instantaneous or an average over a significant time – do you mean instantaneous rate (the tangent slope at a point on a graph of a variable dependent on time); or do you mean the average rate (slope of a line joining two distant points on a possibly non-linear graph)? Expressions such as 'stocking rate (ha^{-1})' are not time-based and should not be called a rate but, in this example, a density.

ratio / quotient: 13 : 3 is a ratio, but 13 / 3 is a quotient with the numerical value 4.33. A quotient can be reduced to a single numerical value (for plotting or calculation) but a ratio always has two numerical values separated by a ':'. A ratio cannot therefore be used easily as an axis on a graph. The two numerical values in a ratio are often integers smaller than 100. Almost always you need 'quotient' but the word 'ratio' is deeply rooted in numerous named quantities (for example Bowen ratio, C : N ratio, Golden Ratio). The two components of a ratio, as with an efficiency (*q. v.*), should have the same physical dimensions, or none. A quotient is not so restricted, it simply indicates division. So if the components have different physical dimensions (Chapter 7) then the quotient will itself have physical dimensions and hence units. 'The C : N ratio was 39:1' (but were the units mass or were they mol?), 'the respiratory quotient was 3.6', but *not* 'the C : N ratio was 39' (missing ':1'), or 'the C / N ratio was 39' (C / N is a quotient).

strain: 'strain' is the result of 'stress': 'stress' is the cause of 'strain'. The neglected polymath Robert Hooke (who had the misfortune to be nearly a contemporary of Isaac Newton) codified the relationship in 1676 when, in an attempt to establish priority, he published *A Decimate of the Centesme of the Inventions I Intend to Publish*. One of these claims was the anagram 'ceiiinossstuu'. An explanation followed in 1679 in *De Potentia Resitutiva or of a Spring:* the anagram is unveiled as 'Ut tensio sic uis [now 'vis']' – 'As the extension so the force' (Gordon 1968). In a standardised form the slope measures the stiffness of the material, and is Young's modulus. Rather surprisingly, however, Young's modulus is stress divided by strain implying that the causal independent axis is the ordinate and the resultant dependent one is the abscissa: the opposite of the present convention. The explanation may be that the engineers' apparatus for stretching wires increases stress until a predetermined increment of strain has been reached. They fix the strain required and measure the stress observed to get it. This does not alter the physics in which the strain is caused by stress.

stress: see *strain* (next above).

variable / variate: statisticians distinguish a mathematical variable, which is a quantity that may take more than one value, from a variate which is a set of observed values of a variable. A graph of a dependent variable will be a line, while the realisation of a variable as a dependent variate will be a series of points scattered about an imaginary trend line. A variable or variate is often wrongly called a parameter (*q.v.*). A variable has variability (it has the potential to vary); a variate shows variation (Webster 2001).

7

Quantitative matters

If you never quote a numerical value, never use units, never use graphs, and never write equations then you can safely skip this chapter.

Numerical values

The meaning and usage of 'number', 'numeral', 'digit', 'value', and 'figure' is not entirely clear.

- 'Number' has the broadest meaning. Illuminating accounts of the relationships between different sorts of number – integer, rational, irrational, imaginary, complex, transcendental – are given in Feynmann, Leighton, & Sands (1963) and in Sondheimer & Rogerson (1981). 'Number' is commonly used as well for counted items: 'The number of sheep', and in a general sense as in 'the numbers show …'.
- 'Value' applies to variables, variates, parameters, and constants. (The proper expression of such a value is considered later in this chapter.) Each has a 'value' that comprises two components: a number *and zero or more* units, as in '24.6 mg'. In most places in this book I use 'numerical value' for the number part of the overall value to emphasise this point, but the simple 'value' is widely used for the numerical component alone.

- 'Numeral' is most often used for Roman quantities (as in dates such as MCCCCLVIII).
- 'Digit' is one of the single Arabic symbols '0' ... '9' corresponding to human fingers or toes. Thus all real numbers can be expressed as a combination of Arabic digits with the symbols '+', '−', '.' and perhaps 'e', or 'E', or '×'.
- 'Figure' is unclear, and has a different and specific use for illustrations, so is best kept for them alone.

In general, one should try to keep the numerical component of a value within the range 0.01 to 1000 by adjusting the multiplier of units: '0.00025 g' is more easily understood as '0.25 mg'. But it may be more important to use the same multiplier than to get the numerical component within a limited range: 'the mean mass of fallen branches was 1.5 kg m^{-2}, while that of flowers was only 0.00025 kg m^{-2}' emphasises the small proportion of flowers, while 'the mean mass of fallen branches was 1.5 kg m^{-2}, and that of flowers was 0.25 mg m^{-2}' obscures this.

In scientific text, integer values 'one' to 'nine' should be spelled out unless they are part of a quantity with units: 'five sheep', but '6 cm s^{-1}'. Larger integers, negative integers, and decimal numbers with an exponent should usually be in digital form: 17, −5, 1.06, 3×10^{27}, −7e−45. Within a sentence or larger chunk of text, however, be consistent. Thus '... the number trapped ranged between 2 and 27'. Sometimes, it may be necessary to use spelled out and symbolic representations of numerical values if two series are to be mixed: 'there were six examples containing 15 animals and twelve containing 17'. Numbers at the start of a sentence (better avoided, particularly if part of a range such as '16 to 73') should be spelled out: 'Fifteen men on the dead man's chest.' Avoid the ambiguity in statements that use 'by' before two numerical values as in 'The mass increased by 3 to 12 g day^{-1}': did the daily range of individuals increase from 3 at the start to 12 at the end, or was the increase of individuals during the period in the range from 9 to 12?

A recent recommendation (Style Manual Committee, Council of Science Editors 2006) is that science editors change to using digits for all numerical values, with some exceptions such as at the start of a sentence.

Report decimal numerical values less than one with a preceding zero: '0.15' not '.15'. A reader of .15 may too easily miss the point (in both senses).

It is difficult to get the meaning from 37984160334 (or even 82400000000): one has to go to the right end then work backwards dividing into groups of three digits (we owe this to taking our system from an Arabic source with its right-to-left convention). Only then can we see whether this is 3 billion, 37 billion or 379 billion ▾. To avoid this difficulty one can insert a ',' or a 'space' between groups of three digits (both before and after the decimal point) so the number becomes 37 984 160 334. An exception is made for numbers in the range 1000 to 9999: the group separator is omitted. 'Space' may be the better separator because ',' is used in many countries to indicate the decimal 'point', but be careful to avoid splitting space-separated numbers over two lines (use 'hard' spaces).

▾ There used to be potential confusion between the USA billion (10^9, a thousand million) and the UK billion (10^{12}, a million million). As monetary inflation rolled on so the budgets of countries (and more recently the deficits of banks and other corporations) moved from hundreds of millions to thousands of millions, and journalists recognised the convenience of the USA billion. That version is now all but universal. The USA trillion (10^{12}) has become common, and even the quadrillion (10^{15}) can sometimes be seen in flight.

Here is a suggestion for debate. Large and small numerical values have conventionally been represented as multiples of a power of 10: for example 3.42×10^{20} and -2.7×10^{-25}. The '×' and the superscript are typographically awkward, and the smaller size of the superscript can make it difficult to read. When computers first became common, programmers were forced to find some way of representing large and small numerical values

> but without using superscripts. They chose, for the example numerical values above, 3.42E20 and −2.7E−25, the 'E' indicating 'exponent', uppercase because only uppercase letters were available at that time. Nowadays better versions are 3.42e20 and −2.7e−25, the half-height 'e' being easier to notice among its taller neighbours than is 'E'. The 'e' version is almost universally understood, and is two characters shorter than the conventional superscript one too. Some journals specifically forbid such usage. Has the time come to accept the 'e' convention in scientific text?

Symbols

Journals differ in their rules about symbols for variables, variates, parameters, and constants (and a few seem to have no rules). The British Standards Institute (2005) specifies italic (sloping) face for variables, variates, parameters, and physical constants (with bold lowercase for **vectors**, and bold uppercase for **MATRICES**), but 'roman' (upright, 'normal') for mathematical constants such as 'π' and 'e'. A subscript or superscript that is an abbreviation, an identifying letter, or a number should be roman (the subscript '3' in 'b_3' for example). Subscripts and superscripts that are variables should be italic. Thus x_n (italic 'n') is the 'n-th' of the 'x's, but x_n (roman 'n') is 'x' in compartment 'n' (the other compartment being, for example, 'p').

It is helpful to represent variables, variates, parameters, and constants by a single uppercase or lowercase Latin or Greek letter, with pre-script or post-script, subscript or superscript qualifiers. You then avoid the ambiguity in 'Mb', for example. Is this 'M' multiplied by 'b' or is it a compound identifier of the single variable 'Monoblock index'? 'Mb' (not italic) can also represent the unit 'mega bit' (perhaps a mistake for 'MB'), or even (in molecular biology) 'mega base-pair', though the context should prevent the last confusion.

Chemical symbols (such as 'Ca'), pre-multipliers and unit symbols (such as 'cm'), and mathematical functions (such as 'tan') and operations (such as '=') should all be roman (upright) face.

Chemical nomenclature (Style Manual Committee, Council of Science Editors 2006), particularly of organic chemicals, and the consequent rules about symbols, are elaborate and mostly beyond the scope of this book. The current practice is to represent proton number as a pre-superscript, for example, ^{14}C, and nucleon number as a pre-subscript, $_6$C. Examples of a common format for organic radioactive tracers are '[^{3}H]arginine', '[6-^{14}C]glucose' and '[U-^{14}C]glucose' (the 'U' indicating uniform labelling). Word-processors may be unable to align superscript and subscript directly above and below one another: if the journal requires this let it arrange it. In text one can nowadays write 'C14', omitting the hyphen that used to be required. Note that the '14' follows the 'C' in this form, but precedes it in '^{14}C'.

The developing symbolisms of genetics and molecular biology (Style Manual Committee, Council of Science Editors 2006) are also beyond the scope of this book.

Units and multipliers

In 1960 the Conférence Générale des Poids et Mesures (CGPM) finally brought together in the Système International d' Unités (SI) the long-continued efforts by several international bodies to standardise a system of units. Most scientific journals now require or, at least, recommend it. The seven basic independent quantities and units are shown in Table 7.1.

Originally most of these quantities were defined in terms of a specific object. For example the metre was defined as 1/10 000 000 of the quarter meridian of the Earth running through Paris, and later as the distance between two marks on a particular bar of Pt–Ir alloy kept at a specified temperature in Sèvres (France). The kilogram standard is a particular block of Pt–Ir alloy, also kept at Sèvres. (The base unit of mass, 'kg', appears to be a multiple of a smaller unit, 'g'. This anomalous decision was made at least partly because the CPGM feared that it would take too long to get acceptance of a totally new name.) As measurement precision increases the definitions are being changed from unique artefacts,

TABLE 7.1 The seven SI (Système Internationale) base units. The steradian as 'solid angle' in the definition of candela, and the radian, are overall dimensionless quotients with the value 1. They were once classed as supplementary base units but are now classed as derived units (Table 7.2)

Name	Unit [dimension]	Symbol	Physical basis
Mass[a]	kilogram [M]	kg	A particular block of Pt–Ir alloy
Length	metre [L]	m	Distance travelled by light in a vacuum during a specified time
Time	second [T]	s	Transition of two hyperfine levels of the ground state of ^{133}Cs
Electric current[b]	ampere [I]	A	Force between two parallel straight conductors
Thermodynamic temperature	kelvin	K	1/273.16 of the thermodynamic temperature of the triple point of water
Amount of substance[b]	mole [N]	mol	Same number of specified entities as there are atoms in 0.012 kg of ^{12}C
Luminous intensity[b]	candela [C]	cd	Radiant intensity of specified frequency in solid angle as energy / solid angle

[a] Depends on a specific artefact.
[b] Mole, ampere, and candela depend on the definition of mass.

which nevertheless do change very slowly, to those based on physical processes which, in principle are invariant and can be consulted anywhere. Time was the first to be defined in this way by the frequency of an atomic process, then length in terms of the speed of light in a vacuum and time. Mass still depends on a particular Pt–Ir block, and electrical current, the chemical mole, and luminous flux depend on mass. There is a proposal to move to a 'new SI' in which all units are defined in terms of universal physical constants: mass (kg) on the Planck constant (h); electric current (A) on elementary charge (e); temperature (K) on the Boltzman constant (k); and chemical amount (mole) on the Avogadro constant (N_A). These proposed changes would be made only when the four constants are known to better precision than the current standards. You can find clear explanations of the intricate details on the website (<www.bipm.org/>: 6 Sept. 2013) of the Bureau

UNITS AND MULTIPLIERS

TABLE 7.2 The 22 SI named derived units. Angles (radian, steradian) are nowadays considered to be derived (they are considered as if they were dimensionless and have the value 1)

Quantity [dimensions]	Name	Symbol	Units	Quantity [dimensions]	Name	Symbol	Units
Plane angle	radian[a]	rad	m m^{-1}	Electric charge [TI]	coulomb	C	s A
Solid angle	steradian[a]	sr	m^2 m^{-2}	Electric potential difference [ML^{2}T^{-3}I^{-1}]	volt	V	kg m^2 s^{-3} A^{-1}
Force [MLT^{-2}]	newton	N	kg m s^{-2}	Resistance, impedance, reactance [ML^{2}T^{-3}I^{-2}]	ohm	Ω	kg m^2 s^{-3} A^{-2}
Pressure, stress [ML^{-1}T^{-2}]	pascal	Pa	kg m^{-1} s^{-2}	Conductance [M^{-1}L^{-2}T^{3}I^2]	siemens	S	kg^{-1} m^{-2} s^3 A^2
Energy, work, heat [ML^{2}T^{-2}]	joule	J	kg m^2 s^{-2}	Capacitance [M^{-1}L^{-2}T^{4}I^2]	farad(ay)	F	kg^{-1} m^{-2} s^4 A^2
Power, radiant flux [ML^{2}T^{-3}]	watt	W	kg m^2 s^{-3}	Inductance [ML^{2}T^{-2}I^{-2}]	henry	H	kg m^2 s^{-2} A^{-2}
Frequency [T^{-1}]	hertz	Hz	s^{-1}	Magnetic flux [ML^{2}T^{-2}I^{-1}]	weber	Wb	kg m^2 s^{-2} A^{-1}
Luminous flux [CSr]	lumen[a]	lm	cd sr	Magnetic field, flux density [MT^{-2}I^{-1}]	tesla	T	kg s^{-2} A^{-1}
Illuminance [L^{-2}CSr]	lux[a]	lx	m^{-2} cd sr	Radioactivity [T^{-1}]	becquerel	Bq	s^{-1}
Temperature	celsius	°C	K−273.16	Absorbed ionising radiation [L^{2}T^{-2}]	gray	Gy	m^2 s^{-2}
Catalytic activity [NT^{-1}]	katal[a]	kat	mol s^{-1}	Equivalent ionising radiation [L^{2}T^{-2}]	sievert	Sv	m^2 s^{-2}

[a] Not a person. All the other derived units in this table are named for a person, but the name of the unit itself has a lowercase initial.

International des Poids et Mesures (BIPM) which is the small permanent secretariat that maintains SI between meetings of the CGPM. Most bioscientists will never need to worry about these details, but journal editors do require that authors use the base units and those derived from them correctly.

There are uncounted derived quantities that combine the base units. Twenty-two of these have officially recognised names and symbols (Table 7.2). Finally there are non-standard units (Table 7.3) ranging from 'officially tolerated' through 'refusing to die' to 'moribund' to 'dead'. There are numerous printed works

TABLE 7.3 Some other derived units, permitted by SI, or in widespread use but discouraged with SI

Quantity	PERMITTED UNIT	Quantity	DISCOURAGED[a] UNIT
Angle	degree (°), arcminute ('), arcsecond (")	Angle	grad (1/400 of a circle)
Mass	atomic mass unit (u or Da), gram (g), metric tonne (t = 1000 kg)	Length	Ångström (A = 0.1 nm)
Length	astronomical unit (AU)	Area	hectare[b] (ha = 10 000 m^2), are (100 m^2), barn (1.0e-28 m^2)
Time	minute (min), hour (h, hr), day (d, dy), year (a, y, yr)	Pressure	bar (100 kPa), millibar = mbar = mb = 0.1 kPa), torr, atm
Volume	litre (l, L)	Velocity	knot (kt)
Work, energy	electron volt (eV)	Work, energy	calorie (cal), (erg)
Sound quotient	bel (B), decibel (dB)	Viscosity	(poise, Stokes)
Information	(bit)	Temperature	degree Fahrenheit (°F)
Information flux	(Baud)	Radioactivity	(Curie, rad, Röntgen, rem)

[a] These DISCOURAGED units are examples only: there are uncounted others. In particular all units derived from the old CGS system (but not cm, g, and s themselves) are incompatible with SI.
[b] The hectare is so widely used (for example in ecology, soil science, and geography) that it is difficult to believe that it will be dropped. The same applies to mbar in meteorology.

with more details, for example McGlashan (1971), Massey (1971), and a lot of websites.

You will also need to use multipliers to bring numerical values into a range (0.01 to 1000) that is easy to grasp. Thus one could select an area in μm^2, in mm^2, in cm^2, in m^2, in hm^2 (ha), or in km^2. These could be appropriate at one end for measurements within cells, and at the other for forests. Table 7.4 lists the SI-sanctioned multipliers. Notice the distinction between 'd' and 'D', 'm' and 'M'. If you need them you may have to explain the multipliers 'a', 'f', 'P', and 'E', though they are becoming commoner as techniques and interest expand to the smaller and larger. You had better

197 UNITS AND MULTIPLIERS

TABLE 7.4 Standard prefixes for units. The International Electrotechnical Commission (IEC) has proposed prefixes for powers of 2 (n in 2^n), but these are not popular. For example, in SI, 1 kb = 1000 bit, and in IEC 1 Kib = 1024 bit

\multicolumn{4}{c	}{Submultiple}	\multicolumn{4}{c}{Multiple}
Value	Symbol	Name	Origin	Value	Symbol	Name	Origin
10e−24	y	yocto	Greek	10e24	Y	yotta	Greek
10e−21	z	zepto	French	10e21	Z	zetta	French
10e−18	a	atto	Da, No[a]	10e18	E	exa	Greek
10e−15	f	femto	Da, No[a]	10e15	P	peta	Greek
10e−12	p	pico	Spanish	10e12	T	tera	Greek
10e−9	n	nano	Latin	10e9	G	giga[b]	Greek
10e−6	μ	micro	Greek	10e6	M	mega	Greek
10e−3	m	milli	Latin	10e3	k	kilo	Greek
0.01	c	Centi	Latin	100	h	hecto	Greek
0.1	d	Deci	Latin	10	da	deca	Greek

[a] Da, No = Danish and Norwegian
[b] Once pronounced as in 'gigantic' but now commonly as in 'giggle'

explain 'y', 'z', 'Z', and 'Y' if you need them. (In mid-2003, in the journal *Science*, I first came across an unexplained use of 'Z'. Laser powers in PW cm^{-2} are now common, EW cm^{-2} are needed for new designs, and there is discussion about the feasibility of ZW cm^{-2} designs.)

Notice the following.

(i) SI specifies that the decimal point should be a dot or full stop or 'point' ('.') or comma (',') on the line. This is a relief as the dot at mid height (·) is a special character, not on keyboards, and hence awkward to type.

(ii) For the minus sign use the special 'minus' character '−' (Unicode 2212), or the 'en' dash, '–', because the hyphen ('-') on the keyboard is not conspicuous enough. (The keyboard '-' is next to the '+', so is often, confusingly, thought of as 'the minus sign'.)

Your word-processor may be persuaded to replace a double hyphen '- -' by the 'en' dash '–'.

(iii) Use only a single pre-multiplier: not '5.6 mµg' but '5.6 ng'. Note that for mass one ignores the fact that the base unit is really 'kg' and treats mass as if the base to which the multiplier is applied were 'g'.

(iv) Units should be singular and without a '.' after them, however large the numerical value. Thus '27.6 g', not '27.6 g.' or '27.6 gr' or '27.6 grs' or '27.6 gram' or '27.6 grams' or '27.6 grammes' or any other variant. Similarly 'mole' is abbreviated to 'mol', not 'mols' or 'moles' or 'Mol' or 'Mol.'.

(v) If an index is used then it applies to the pre-multiplier as well as to the unit: cm^3 is $(cm)^3$ not $c(m^3)$.

(vi) There is no space between a pre-multiplier and the unit it applies to, but there *is* a space between different units, and between a number and its unit(s). Exceptions are in angles as in 'latitude 6° 7' 37" W' (and in longitude). Some journals, however, allow multiple units to be separated by a centred dot (*not* a full stop, on the line): 'm·s^{-1}'. Thus one can distinguish the velocity 'm s^{-1}' from the inverse time 'ms^{-1}' (per millisecond). Symbols for percent (%) and for temperature (K, °C) should be treated as units:

> not '... 27microgs per ml. (2.7 % solution) boiled at 37 degC. for 30 mins.'
> but '... 27 µg ml^{-1} (a 2.7 % solution) boiled at 37 °C for 30 min.

(vii) Word-processors take licence to break a line wherever there is a space ' ' or a hyphen '-' (which you may have intended as a minus sign though it should be a '−' or 'en' dash '– '). To avoid automatic breaks in the middle of compound units, or separation of the number from its units, you can use the 'non-breaking' ('hard') versions of ' ' and '-' available with word-processors. These characters appear identical to the ordinary versions but prevent the word-processor breaking a line at them. Thus, with non-breaking spaces (and hyphens) in use, we find that 25.7 mg m^{-2} yr^{-1} is treated as a single uninterrupted word that will be kept together as a unit, while with soft ' ' and '-' we get 25.7 mg m^{-2} yr^{-1}.

(viii) The SI multipliers cannot be used for quantities with no units such as '357 000 cells'. If you need a short form, on a graph axis for example, use Cells × 10^{-5} (or 'Cells × 10e−5') and the corresponding number 3.57. Note that it is the original cell number that has to be multiplied by the factor to give the quoted or graphed number.

(ix) Some latitude may be allowed for particular units. For 'ppm' and 'ppb', always specify the basis, for example volume as 'vpm' or 'ppmv'. Some journals take a hard line and require all non-SI units to be spelled in full. They may forbid p.p.m.v. and ppmv for example, except for the chemical shift in nuclear magnetic resonance spectroscopy. This hard line is particularly troublesome with time units. Some journals do tolerate a variety of abbreviations: the abbreviation for day may be 'd' or 'dy'; hour may be 'h' or 'hr', and year may be 'y', 'yr', or 'a' (for 'annum'). There are several sorts of year: calendar (civil), several astronomical years (sidereal, tropical, and anomalistic), and 364.25 dy. The unit of volume should be 'dm^3' but in concentrations 'l^{-1}' or, to avoid confusion, small capital 'L^{-1}' may be tolerated. The metric tonne (symbol 't') is widely accepted in place of the formal Mg, though if 'metric' is omitted then confusion of this ton is possible with the imperial ton = USA long ton (2240 lb = 1.02 Mg) and the USA short ton (2000 lb = 0.91 Mg). If the rules for separating the name of a variable from its units are followed then Mg (the element magnesium) will not be confused with Mg (mega-gram).

(x) Other units of time are uncertain too. The 'day' may be exactly 86 400 s or it may be a calendar day. The 'month' may be four weeks (28 days) or a named calendar month. These units are uncommon in scientific writing except in the context of 'The experiment lasted 21 days and was repeated every six (calendar) months for ...'.

(xi) Four more non-SI units flourish in spite of discouragement. The 'Å' is named after the Swedish physicist A. J. Ångström (1814–74; pronounced 'awng-strerm') and is 10^{-10} m (0.1 nm). It is particularly useful for distances between atoms, and this may partly account for its continued popularity. The hectare (ha) is also officially discouraged, and so is its base, the 'are' (100 m^2), though the 'ha' will probably survive because it is a single entity (a named derived unit) of a convenient size that SI cannot easily replace. The 'bar' unit of pressure (particularly as 'mbar') is widely used in meteorology, and the 'knot' in maritime matters

The main features of SI are clear, but the finer details are a quasi-legal playground. Follow the main rules listed above but do not fret over minutiae.

Dimensions and equations

Dimensions must balance

The '=' in an equation is like the knife-edge of a pre-electronic era balance: the things in the pan on one side are 'equal to' the things in the pan on the other side. The commonest published equations have one element on the left and an expression, often with several elements, on the right. A simple example is the straight line

$$y = bx + a \qquad [7.1]$$

which is said to be 'explicit for y' meaning that if you are given numerical values for b, x, and a then you can calculate y. This equation is mathematically unobjectionable, in that one cannot say that there is anything wrong with it as it stands. Now suppose that the same straight line relationship is claimed to describe the way that the height (h) of the main shoot of a climbing bean plant increases with time (t). Change the symbols but not the structure to get

$$h = ut + v. \qquad [7.2]$$

What can we say about the parameters u and v? (See 'Frequently misused words' in Chapter 6 for the difference between a variable and a parameter.) As h is a height then v must also be a height, and so must ut: you can add or subtract heights but you cannot sensibly add a height to, for example, a mass. So the parameter u must have the nature of a height divided by a time, so that when it is multiplied by the time, t, the result is a height. Each variable and parameter in equation [7.2] has physical dimensions. The dimensions common in mechanics are mass, length, and time, usually represented by [M], [L], and [T]; and [n] can be used for a physically dimensionless numerical value. (Square brackets '[]' are used here to enclose both equation identifiers and dimensions.) In [7.2], h is a length, so we have just inferred the dimensions v [L] and u [LT^{-1}]. The whole equation can be written in terms of dimensions as

$$[L] = [LT^{-1}] [T] + [L]. \qquad [7.3]$$

The essential feature of this is that equations claimed to describe the behaviour of the real world must balance EACH of their physical dimensions separately. This is a powerful aid in checking the validity of your own equations and detecting the frequent inadequacies, or even errors, in other authors' publications.▼ How often have you seen a table containing parameter numerical values from a best-fit linear relation, but without the essential accompanying units, such as 'mm hr^{-1}' for u in [7.2] above?

▼ Any equation containing a variable representing pH should be viewed sceptically because the definition of pH is dimensional nonsense, requiring as it does the impossible operation of taking the logarithm (an operation that can be applied only to a number) of a concentration (a physical variable).

Combining units and dimensions

In practice, you may find it easier or more convenient to substitute dimensions with the corresponding units. For example the bean growth equation [7.2] can be restated with SI base units as

$$[m] = [m\ s^{-1}] [s] + [m]. \qquad [7.4]$$

As with dimensions, the equation must balance for each unit separately: for [m] and [s] in [7.4]. If these units are inconvenient then use a suitable pre-multiplier, for example

$$[mm] = [mm\ hr^{-1}] [hr] + [mm]. \qquad [7.5]$$

All that is necessary is that you use the same pre-multiplier wherever a particular base unit appears. When you come to putting data into the equation then you must use the same pre-multiplier of the datum as is in the dimension equation: height in mm, time as hour, in the example.

If you do use units instead of dimensions then be careful to avoid the temptation to treat [7.4] as if it were an algebraic statement that reduces to the absurd **[m]** = 2 **[m]**. For this reason I show the units in bold face.

'Equations' that are not: when dimensions do not balance

Now consider another (unlikely) equation

$$y + \ln(y) = x / \exp(x). \quad [7.6]$$

This equation is 'implicit for y': for every numerical value of x there is only one of y that satisfies [7.6], but you cannot calculate it directly. ▾ The graph of [7.6] is shown in Figure 7.1. How does this fare if it is claimed to apply to a physical situation rather than just to numerical values? Not well. Suppose for example that y represents a concentration $[ML^{-3}]$ and x a length $[L]$. Then what does '$\ln(y)$' mean? We can take the logarithm of a *number* but not of a variable, such as concentration, which represents some *physical*

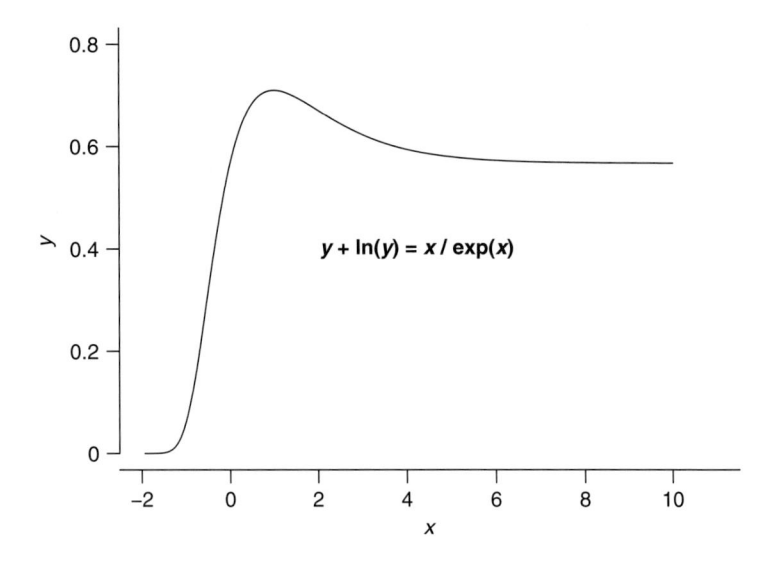

FIGURE 7.1 The function $y + \ln(y) = x / \exp(x)$.

quantity. The expression 'ln (y)' is then meaningless. Can this defect be corrected? Yes, easily, and at the same time let us correct the similar error on the right-hand side:

$$y - \ln(ry) = x / \exp(sx). \tag{7.7}$$

The parameter r must have the character of $1/y$ (a reciprocal concentration $[M^{-1}L^3]$) and s must have the character of $1/x$, i.e. a reciprocal length $[L^{-1}]$. Is this sufficient? No, because [7.7] requires that a dimensionless number, $\ln(ry)$, be subtracted from a concentration, y. Correct this by adding the parameter a to give the dimensionless $ay - \ln(ry)$. The right-hand side is acceptable in itself because a length is to be divided by a number giving a length, but it has the dimension $[L]$, while the left-hand side is $[n]$. To make the whole equation consistent in the physical world we need something like

$$a\, y - \ln(r\, y) = b\,[x / \exp(s\, x)] \tag{7.8}$$

where both sides are dimensionless, and the new parameters are $a\ [M^{-1}L^3]$ and $b\ [L^{-1}]$. The dimension equation is now

$$[M^{-1}L^3][ML^{-3}] - \ln([M^{-1}L^3][ML^{-3}]) = [L^{-1}][L] / \exp([L^{-1}][L]). \tag{7.9}$$

If the numerical values of y are in g dm^{-3} then the parameter a may be 1.0 dm^3 g^{-1}, though it could be used to apply any factor to the numbers in the equation. Some scientists object to introducing a scaling parameter such as a, often with the numerical value 1.0, solely to make the equation balance dimensionally. I believe they are mistaken: dimensional balance is such a powerful aid in detecting errors that one should always aim to achieve it.

▼ An equation implicit for y cannot be manipulated to the form $y = f(x)$, yet for each numerical value of x there is only one numerical value (or, in general, a specific number of numerical values) of y that satisfies the equation. Given a numerical value for y, how does one find that of x? First, move everything to the left-hand side to give:

$$y - \ln(y) - x / \exp(x) = 0$$

Next, take a valid numerical value for y (between 0.0 and 0.7 in the example, Figure 7.1) and calculate the numerical value of the left hand side, using a trial valid numerical value of x. Suppose the result is not 0 but a positive numerical value. Now halve the numerical value of x and recalculate. Suppose the numerical value is now negative. The true numerical value of x that will give 0 for the left-hand side must lie somewhere between the first and second guesses. Try half way between them. If the result is positive then discard the first guess (that also resulted in a positive answer); if negative discard the second (negative) guess. The range around the true numerical value is now reduced. Continue in this way, trying numerical values halfway between ones that are known to give results with opposite sign until the result is acceptably close to zero. This whole process – finding the zero of a function by bisection – can be efficiently coded for a computer. Solutions of [7.6] for 1000 numerical values of x were produced in less than a millisecond (Figure 7.1). Often there are more efficient methods than that outlined here, but all methods depend on getting two initial numerical values of y that give results that bracket zero. An excellent account of such methods, with programs, is given in Press *et al.* (1989).

Sometimes an equation presents itself in nonsense form but can be rearranged without extra parameters to make sense. For example, from a respected limnology text we have a description of the depth to which light penetrates in the water of a lake or sea

$$Z = [\ln (I_0) - \ln (I_k)] / \varepsilon \qquad [7.10]$$

where Z [L] is depth, I_0 and I_k [ML^2T^{-3}] are radiant flux incident at the surface (0) and at a depth (k) defined in a way that need not concern us here, and ε [L^{-1}] is the vertical attenuation coefficient. As I represents a variable with physical existence it is impossible to take a logarithm of it. But we can rearrange [7.10] to give

$$Z = \ln (I_0 / I_k) / \varepsilon \qquad [7.11]$$

which is dimensionally correct.

Using dimensions to decide what to do: an example

Suppose that you want to draw a random sample from the (assumed 'normal') distribution of the mass of pumpkins, which you know to have a mean of 24.0 kg and a standard deviation (SD) of 6.0 kg. Your computer will provide you with a standard 'normal' distribution with a mean of 0.0 and a standard deviation of 1.0. These are pure numbers – no units. How do we get from this to the required distribution of masses (Figure 7.2)? Two operations are necessary: the mean must be shifted to the right (by adding the mean mass) and the spread must be increased by expanding (multiplying) by the mass SD. In which order should these operations be applied? A little thought will give the answer, but let us apply a dimensional argument. Suppose we begin with the shift of the mean. This requires adding 24.0 kg to the pure number 0.0 – a dimensionally forbidden operation. Even if that were allowed we

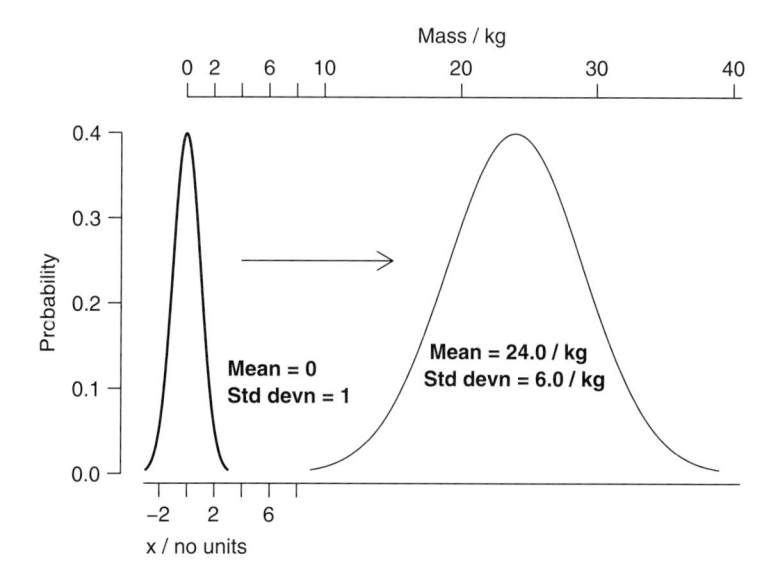

FIGURE 7.2 *Left, lower axis*. Standard 'normal' distribution, mean = 0, SD = 1. To be transformed to (*right, upper axis*) probability of mass with mean = 24.0 kg, SD = 6.0 kg. This requires a horizontal shift of the mean, and an expansion of the SD. In which order should these two operations be made? The lower and upper axes have the same *numerical* scale, but the lower is of pure numbers while the upper has units (kg).

would then have to multiply a mass by another mass (the SD) giving a result with dimensions [M^2] not the required [M]. Instead we must first perform the multiplication of the pure number 1.0 by the SD, to give a mass, and then add the mean (a mass too).

Expressing the name and value of a variable

This section concerns variables, variates, parameters, and constants that have a physical existence (for example, a rate) not just a pure number. The value of such a physical quantity has two components: its numerical size, and its units. For example, in [7.2] describing the growth in height of bean plants we may have:

h = 126 cm

which can be rearranged to give the dimensionless version:

h / cm = 126

The '/' indicates the operation of division, so the physical quantity 'height, h', which has dimension [L], is divided by the chosen unit 'cm' to give the pure number '126'.

In a table of the results there would be a column (or row) containing the heights, and one cell would contain the dimensionless (pure) number '126'. The title (heading) of the column (or row) would then be 'Height / cm'. If there were a column recording the rate of growth it might have the title 'Rate of growth / mm hr^{-1}'. The slash (virgule, '/') is used to divide the name (or symbol representing the name) of the physical quantity by the units, individual units using index notation such as 'hr^{-1}' where division is needed. Nothing but units, pre-multipliers, and indices appears after the slash. (Some journals allow you to extend the index notation to basis items, as in 'mg $plant^{-1}$'; others forbid this and require 'mg per plant'.) This dimensionless convention separates the name of the physical quantity cleanly from the units, and is consistent with pure numerical values in the cells of a table. It also allows one to have a column (or row) with a transformed title such as, for example, 'Log$_{10}$ (rate of growth / mm hr^{-1})'. This is consistent with the fact that one can take logarithms of pure numbers ('rate of growth / mm hr^{-1}') but not of quantities with physical units. Thus 'Log$_{10}$ (rate of growth)' is nonsense because the numerical value depends on what units are used.

The same arguments apply to the labels on graph axes: 'Rate of growth / mm hr^{-1}' defines dimensionless numerical values, and it is these numerical values that appear by the tick marks on the axes and that are used to place data points. The same applies when an axis is to be transformed: the transform can be applied only to dimensionless numerical values.

For later use I will call this the 'slash convention'.

In bioscience research this consistent convention is often, unfortunately, ignored. Many journals use a convention in which the name of the physical quantity is followed by the units in parentheses, for example 'Rate of growth (mm hr^{-1})'. I will call this the 'parenthesis convention'. Most of the time this should cause no difficulty, though it clashes with other conventions which imply that this is either the product 'Rate of growth times (mm hr^{-1})' or 'Rate of growth is a function of (mm hr^{-1})'. But it is frequently abused in two ways.

First, attempts to apply a transform run into the logical problems already mentioned. These difficulties are considered further in Chapter 10.

Second, the name of the physical quantity is often inaccurate and requires that parts of the name be transferred into the units. For example

NO_3^- uptake (mg N hr^{-1})

claims that the variable is 'uptake', which would be an amount, but the unit 'hr^{-1}' implies that the variable is actually a rate. The units begin with the 'mg', but what is the 'N'? Surely not the force unit 'Newton'? Perhaps it stands for nitrogen? But that is part of the name of the variable: that name should be 'NO_3^--N', not just 'NO_3^-'. One guesses that this label was intended to mean:

Rate of uptake of NO_3^--N (mg hr^{-1}).

Or, better in the slash convention, as

Rate of uptake of NO_3^--N / mg hr^{-1}.

Notice that the name has the physical nature of the variable first: this is a rate. Then the rate is qualified: it is an uptake. Finally uptake itself is qualified: it is of NO_3-N$^-$. 'Rate of NO_3^--N uptake' is also acceptable and is shorter – an important advantage for a label on a graph axis.

Many journals do allow this confusing mixing of name and units. It can, and should, always be avoided.

QUESTIONS:
- Have you ever used dimensions?
- If not, are there parts of your work or writing in which they might have been useful?

More about dimensions

Dimensions of the coefficients in a polynomial

It is unlikely that an analysis of processes would produce [7.6]. But equally implausible equations with no physical or chemical justification are in routine use for the description of trends. Suppose that a trend has two inflections (it is 'S'-shaped) and you have used an off-the-shelf statistical package to fit a third-order polynomial (to get a reasonable fit, the polynomial must be of order one more than the number of inflections) to

$$y = b_3 t^3 + b_2 t^2 + b_1 t + a, \qquad [7.12]$$

where the variables are y [M] and t [T], the age of an organism being used to estimate the mass.

What are the dimensions of the coefficients (parameters)? The first is easy: a [M] which has a simple geometric meaning as the intercept on the ordinate at $t = 0$. The second b_1 [MT^{-1}] is a slope, mass per unit time. The other two parameters are b_2 [MT^{-2}] and b_3 [MT^{-3}], which have no simple interpretation and in which even the concept of a square or cubic time is obscure. Polynomials can help to guide the eye but rarely lead to understanding.

Dimensions other than [M], [L], [T]

One can use the derivation of the variables common in mechanics to establish their dimensions. Consider pressure [??], defined as force [F] / area [L^{-2}]. Force itself is mass [M] times acceleration; acceleration is rate of change [T^{-1}] of velocity; and velocity is rate of change [T^{-1}] of distance [L]. Together these give the dimensions of pressure as:

[??] = [MLT^{-2}] [L^{-2}] = [ML^{-1}T^{-2}].

Another simple example is Einstein's equation $E = mc^2$, where E is energy (force × distance), m is mass, and c is the velocity of light in a vacuum. In dimensional terms this claims, correctly, that

[ML^{2}T^{-2}] = [M] [L^{2}T^{-2}].

In other fields other dimensions may be needed. 'Muchness' of a chemical substance can be expressed as a mass or as the number of particles (for example atoms, ions, molecules, photons). Suitable units may be kg and mol. To make plain that an equation holds for number of particles rather than for mass, a dimension [N] may be defined. Where heat is concerned then temperature [θ] is needed, and electric charge [Q] enters for electromagnetic problems. But one need not be restricted to [M], [L], [T], [N], [θ], and [Q]. If an equation were explicit for force, and at least one term on the right hand side involved force too then it may simplify matters to make force a dimension. The equation for the difference in pressure between the inside and outside of a bubble in water is

$$\Delta P = 2s / r, \qquad [7.13]$$

where ΔP is the difference in pressure, s is the surface tension, and r is the radius of the bubble. Pressure is force / area, and surface tension is force / distance. With [M], [L], and [T] as dimensions we get

$$[ML^{-1}T^{-2}] = [MT^{-2}][L^{-1}], \qquad [7.14]$$

but with force [F] as a dimension we can simplify to only two dimensions:

$$[FL^{-2}] = [FL^{-1}][L^{-1}]. \qquad [7.15]$$

Remember though that the more dimensions there are to be balanced then the greater is one's chance of discovering mistakes.

This example illustrates the principle: anything that you wish to keep track of separately within a particular equation can be made a dimension, and the equation must balance for each dimension separately.

Inferring the dimensions of a parameter

Consider Fick's equation for diffusion in one dimension:

$$F = -D \, A \, \partial C / \partial x \qquad [7.16]$$

in which F is the diffusive flux (amount per unit time, $[MT^{-1}]$), D is the diffusion coefficient, A $[L^2]$ is area, C $[ML^{-3}]$ is concentration, and x $[L]$ is distance. What are the dimensions of the diffusion coefficient D? Rearrange [7.16] to be explicit for D

$$D = -F / (A \, \partial C / \partial x). \qquad [7.17]$$

The dimensions of (the right side of) this equation are

$$[M][T^{-1}] / ([L^2] \, [ML^{-3}][L^{-1}]) = [L^2 T^{-1}] \qquad [7.18]$$

so units such as m^2 s^{-1} or cm^2 hr^{-1} may be suitable for D. For diffusion of many gases in water $D \approx$ 1e–5 cm^2 s^{-1}. Over cellular distances, up to about 1 μm, diffusion is effective in transporting solutes, but received wisdom is that for larger distances it is inefficient and mass flow is necessary. For times from 1 s to 1 day this is true. But 1e–5 cm^2 s^{-1} ≡ 315 cm^2 yr^{-1}. If hundreds or thousands of years are available then diffusion can be an effective process (Clymo & Williams 2012).

Does the value of D depend on whether mass or chemical measure of amount of substance is used? In this example, no, because the [M] dimension does not appear in D. Of course the dimension of amount (mass or mol) in the flux F, must be the same as that used in the concentration C.

It is even possible, but rare, to use dimensional analysis to derive the powers to which variables must appear in an equation while knowing little about the physical processes; Massey (1971) gives a lucid account of this and other aspects of dimensional analysis.

8

Managing error

Most scientists work at the intersection of three processes (the hatched area in Figure 8.1): (1) specifying what question to ask of Nature; (2) expressing the question as a model (often mathematically even if vaguely as, for example, 'Is there a relation between variables x and y?'); and (3) collecting and analysing data from a survey or experiment.

It is easy to ask the wrong question or to specify the wrong model. A plant physiologist observed that the kinetics of uptake of nitrate from solution by the roots of young barley plants resembled the kinetics of enzyme action and asked 'What is the Michaelis constant of the enzyme?' But this was a blind alley: the kinetics he observed were overwhelmingly the result of diffusion through the unstirred layer around the roots. Even when he had recognised this he specified an incorrect mathematical model, though he got close agreement to it with his data. Asking the right question and specifying it in the right form for testing are at the core of advance in understanding. They are specific to the particular problem though and are therefore outside the scope of this book.▼ But the analysis of data (Figure 8.1), and the sorts of error we need to recognise, are within our scope.

▼ It has been argued (Ioannidis 2005) that in the medical field a majority of the relationships that investigators claim exist are

212   MANAGING ERROR

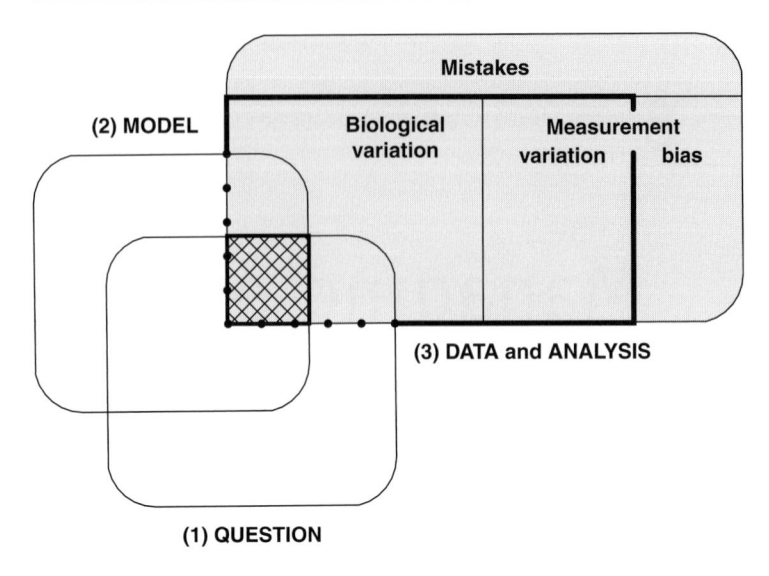

FIGURE 8.1 The place of 'error' of all kinds in scientific work. A particular piece of work will involve (1) one of innumerable questions, (2) a 'model' – a hypothesis, vague or precise, about what may be happening – chosen from innumerable possible models, and (3) data and analysis to test the model. The hatched intersection of these three is the particular piece of work. The analysis in (3) involves three sorts of error: mistakes, bias, and biological and measurement variation. Mistakes and bias must be sought and corrected by the scientist at the bench or in the field. The thick line surrounding the two sorts of variation is the province of statistics.

incorrect. Part of the explanation may be that only the apparently statistically 'significant' results get published, and so, where there is no real effect, the published results are just the fringe of possible results beginning at the conventional '$P \leq 0.05$, 1 in 20' acceptance value. If 100 investigations are made then about five will be judged publishable and report an effect even though in reality the effect is tiny or non-existent.

The sorts of error

There is more than one way of categorising errors, but here I recognise three sorts: 'mistakes', 'bias', and 'imprecision'. You should strive to avoid mistakes, and to minimise bias, but imprecision is unavoidable (though it may be reducible). To assess the amount of bias and of imprecision is an honourable – a necessary – aim.

Mistakes

Many (most?) scientists have made mistakes by factors of ten or a thousand in a calculation, or have reported µmol l^{-1} when they intended µmol ml^{-1} or mmol l^{-1}. Such large mistakes are often obvious enough but smaller ones do not advertise themselves. A few years ago a spacecraft crashed on Mars because a contractor used imperial yards instead of metric metres which resulted in plausible, but disastrous, numerical values. Misreading a graduated scale as 31 when it is 37 amid numerical values that range between 10 and 80 will not announce itself and, if it is the only mistake in a set of 100 measurements, will probably never be detected (and may not matter).

These are human mistakes. There may also be errors in making measurements. Instruments, particularly home-made ones, can give sporadic erroneous numerical values (less often now than when electronics were less reliable).

And mistakes can occur in the material that is being measured: a batch of nominally white tulips might contain the occasional red one; a batch of genetically uniform barley seeds turns out to contain two wild oat seeds.

What can one do to minimise at least the human mistakes?

- First, be aware that mistakes are more likely if you are tired or hungry or in hurry or distracted or otherwise uncomfortable, or using an instrument that is difficult to read because, for example, of poor design, poor lighting, or bad weather. Making yourself as comfortable as possible is a duty.
- Second, when checking be aware that most people tend to expect things to be correct and so may, without intending to, be simply going through the motions of checking. Workers checking the dimensions of nuclear fuel pellets, already checked by two

high-precision automatic machines of fail-safe design, were poorly supervised (mainly because mistakes were so rare). The workers took to copying previous results, with immense financial consequences for their employer: when the practice was discovered the customer for whom the pellets were being made rescinded the contract worth several billion pounds. The report on these practices does not record the error rate in manual inspection, but one suspects that a tiny rate, coupled with the lack of reward for diligence, the perceived unimportance of the task, the sustained and fairly high degree of manual dexterity required in difficult conditions, the lack of automatic entry of results, and the ease of falsification were main causes. When my children were young and needed pocket money I used to get them to help me to check data entries. They were paid at a basic rate with an encouraging supplement for every mistake they found. This had the secondary effect of making me more careful when entering the data.

- Third, concentrate on the matter in hand and minimise distractions. It may be pleasant to have music playing – there are those who maintain that they cannot work without it, or that they work better with it – but is this distraction really without risk?
- Lastly, if your skills allow, get a computer to scan your entered data looking not only for illegal characters, but also for implausible (possibly out of range) numerical values and for impossible combinations of data.

One may strive to avoid all mistakes, but can one succeed? In the long run, probably not. As an example, during the transfer of the printed version of the *Oxford English Dictionary* to electronic form ('keyboarding') the error rate was about 1 in 2000 characters over the 350 million total even with the large element of self-checking by the internal consistency in spelling. A large majority of the errors were human mistakes.

Imprecision and bias

Imagine an archery target (or a dartboard). An archer afflicted with a squint shoots several times aiming at the centre bull's-eye.

THE SORTS OF ERROR

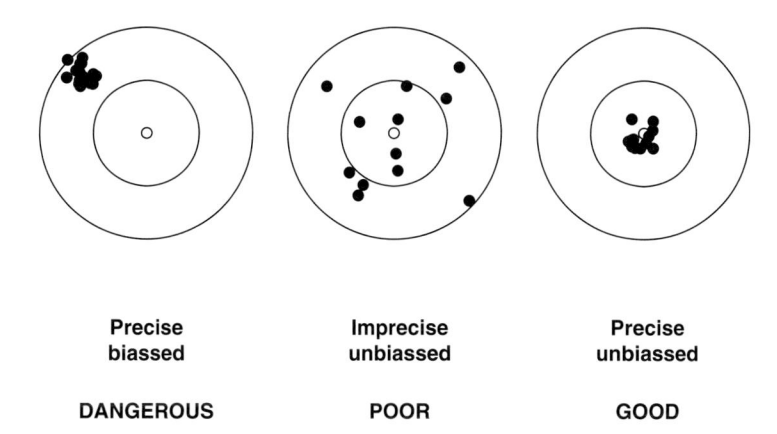

Precise biassed	Imprecise unbiassed	Precise unbiassed
DANGEROUS	**POOR**	**GOOD**

FIGURE 8.2 Precision and bias. *Left*: High precision and high bias (a dangerous combination). *Centre*: low precision and low bias (unsatisfactory but rather obviously so). Right: high precision and low bias (what one would seek).

The arrows all lodge close to one another (Figure 8.2, left) but in the outer circle. Because the arrows are close together the archer's technique is highly reproducible, highly precise – it has small imprecision – but because the mean position of the arrows is a long way from the point aimed at the archer's technique is strongly biassed. Bias is a fixed error in a particular direction (an offset) whereas imprecision has no fixed direction and its magnitude often follows a 'normal' (see later) distribution. The combination of high precision with strong bias is particularly dangerous as the unaware investigator may think that high precision is the only measure of reliability.

A second archer with good eyesight but affected by a tremor in his arms now shoots several times. The arrows lodge apparently randomly in all parts of the target (Figure 8.2, centre) but when their average is calculated (not a usual practice in archery) it is found to be in the bull's-eye. This second archer's technique has small bias but low precision (high imprecision). This result is no more satisfactory than that of the first archer, though it is less dangerously poor because its defects are immediately obvious.

A third archer produces the grouping at the right of Figure 8.2 with small bias and small imprecision. This is clearly what we want.

In a costless world one should aim to eliminate bias. (Many authors use 'accuracy', or more exactly '*in*accuracy', as a synonym for 'bias'.) If it remains large there is no point in going to a lot of trouble to reduce imprecision.

How can one assess bias? ▾ Not easily. If there is an object with a known value for the variable of interest – a standard – and a method that purports to be able to measure the value independently, then the deviation of the mean of repeated measurements from the known value gives the bias. Usually, however, the measurement method 'floats' and the standard is used to anchor the method to zero bias. If there is no standard one may have to use two totally independent methods differing in the principle they rely on, the equipment they need, the person making the measurement, and even in the place the measurement is made. ▪ If two such methods agree to better than their imprecision then one assumes that bias is small. If they disagree then one can review each method seeking hitherto unsuspected sources of bias, and one can use a third method in the hope that this will agree with one or other of the first two and thus direct attention to where the bias may be.

▾ The first recorded example (Alder 2002) of the recognition of bias was after the expedition (1792–99) to measure the length of the quarter meridian of the Earth (which at that time ran through Paris) 1/10 000 000 of which was to define the metre – a plan that we now recognise to have been unsound in principle. Connected triangles with sides several kilometres long were surveyed along the meridian from Dunkerque to Barcelona, and the latitude of the end points was measured by the positions of stars. For both purposes the surveyors, Delambre and Méchain, used a new instrument: Borda's repeating circle. This allowed repeated alternating readings of the direction of two objects to be added together mechanically thus producing a 'better' overall result than a single reading. Many sets of such multiple readings were made at each station and the reproducibility of these and, for triangles, the closeness of the sum of the angles to 180°, was used to assess the exactness of the observations. The new instrument was a great success and allowed Delambre working in the northern part of the meridian and Méchain in the southern part to make observations of hitherto unheard of exactness. In his latitude measurements at Barcelona, Méchain measured six stars from two sites close together and discovered a worryingly large difference in the average of all the measurements at the two sites – a worry that seriously disturbed him for the rest of his life. Not until 1828 did Nicollet explain the discrepancy. He suggested that after thousands of rotations the bearings of the repeating circle had worn slightly so that the circle could rock imperceptibly and measurements to the north of the zenith would be biased when compared with those to the south. When he recalculated Méchain's data in two groups (the four stars to the north of the zenith and the two to the south) the means were indeed worryingly different, but the mean of these two means (rather than the mean for all six stars) was extraordinarily close – about 0.25" of arc – to the same value at both sites (allowing for their small separation measured on the ground). From this emerged the distinction between imprecision and bias.

■ Global average temperature may be measured by several methods. A few years ago measurements at the Earth's surface were trending upwards while those made from satellites were trending downwards. The satellite calculations used the height of the satellite above the Earth's surface, but decay in the orbit of the satellites, so that the height above the surface was decreasing, had been underestimated. When it was corrected the trends in the two methods became similar.

What about imprecision? Here at last we are on fairly firm ground. Standard error (not standard deviation; see later) is a measure of imprecision: the bigger it is then the larger the uncertainty.

Many statistical techniques are available for handling imprecision. Computers have made these easy to use though no more easy to understand, and for these reasons most bioscientists tend to concentrate on managing imprecision and wholly ignore bias – which may be much larger than imprecision.

Measurement error, material variation and error, and randomisation

These errors arise from the variation one finds in repeated measurements of the same object, and 'natural' (inevitable and unavoidable) variation in biological material, its history and environment. They were shown in Figure 8.1 in the area surrounded by a thick line. Most bioscientists look on random variation in repeated measurements of the same object as something they want to reduce sufficiently that differences as a result of treatments are clearly recognisable. They are happy to call such variation measurement error.▼ Those bioscientists who have replicated an experimental treatment recognise that variation in the biological material (and the environment, but I ignore that here ■) also tends to obscure effects in the same way. They consider this too as 'error', and use the same statistical methods to assess it as they apply to measurement error. But to other bioscientists, interested in evolution (for example) this variation in the organisms is the focus of their attention: to them it is the opposite of 'error'. Here though I consider only the first point of view.

> ▼ Measurement error comes partly from equipment and methods, and partly from variations in human skill and diligence (if these are involved). These are not the 'mistakes' already mentioned but variation that appears unavoidable when a procedure is repeated routinely. It is well established that the

> variation among replicates is large when a method is first used but then falls with familiarity to a steady(ish) lower value. It rises immediately after a holiday, or even a minor change in the procedure, or when an operator is worried (threat of redundancy, financial or other personal problems) or almost any disturbance (Young 1949).

> ▪ Fisher, who had experience of the problems in agricultural experiments, introduced 'blocks' and 'blocking' to take account of environmental variation.

Here is an example. An undergraduate student wanted to know whether the activity of an enzyme in pea plants was affected by light flux. He designed and carried out an experiment (ignoring the instructions to seek advice about the design of the experiment before carrying it out). Only afterwards did he ask about what statistical test to use. He had germinated three pea plants in each of two pots, then thinned to one plant in each pot. He had put both pots in a glasshouse and shaded one pot but not the other. After three weeks he removed the plant from each pot, made an extract of each, and then made 10 assays of enzyme activity on each extract. One assay had failed for technical reasons (he dropped the tube). He sought assurance that this imbalance would not prevent some sort of t-test. Of course it is no obstacle, but a more serious problem emerges. The design gives a good estimate of the imprecision of the measurement technique. But there was no replication of the biological material or of the treatment – only one plant had been sampled in light and one in shade. This mistake is often called 'pseudo-replication'. One must make sure to include both measurement error *and* material and environmental error in the overall estimate of imprecision.

If the student had grown 10 plants, each in its own pot, shading five pots chosen and arranged on the bench at random, then made two measurements on each plant he could have made an analysis of variance and got almost as good an estimate of measurement error as he actually had *and* been able to estimate material and environmental error too. And all with negligible extra work.

The need for random allocation of light and shade treatments may be obvious, but less so is another feature of the original design. The student, being of orderly mind, had made both extracts, then made all 10 measurements on the light extract before making 9 (and dropping a tube) on the shaded one. It was only later that he discovered that the extracts deteriorated with age so there was a time-dependent bias in favour of the light treatment values. He should have taken samples for measurement at random (or at least alternately) from the two treatments.

How often have you seen a published graph of the effects of temperature on a single batch of organisms, and discovered (or been left to suspect) that the temperatures were applied *systematically* (from lowest to highest or the reverse). Thus later treatments included all the residual effects of earlier ones with what bias one can only guess. Again one needs at least that the return series should have been followed and that it shows no systematic difference from the first. Better is a random application of different temperatures, or, better still, putting different batches in each temperature.

Lastly, recognise that a random process does not necessarily produce random numbers. Even the best process will, sooner or later, produce regularities such as 1, 2, 3, 4, 5, 6. The shorter the series the greater the probability of obvious regularity. What one really needs is a set of numbers that passes all the possible tests for non-randomness. That is a counsel of perfection though. In practice, do not be afraid to ignore sets that show obvious regularity.

QUESTIONS:
- Can you list the possible sources of bias in a piece of your own work or planned work?
- How would you seek to measure them?

Measures of dispersion and imprecision

There can be few bioscientists nowadays who have not been exposed to statistical instruction, yet many have resisted infection surprisingly well. Defects in these matters are among the commonest criticisms by referees. Here, therefore, are some basics. You can find unusually clear explanations of basic statistical matters in Webster (1997, 2001, 2002).

Suppose you have a set of numerical values of a single variable. On one hand you might need to characterise, that is to describe the distribution of values (for example, of a measurement method), or on the other hand you might wish to show how (im)precisely a mean value is known, so that you can compare two or more mean values. The calculations are similar, though not identical, but the purposes are entirely different.

Dispersion and describing: standard deviation (SD) and quantiles

Consider first the *description* of the distribution of measured numerical values of a variable. To illustrate, Figure 8.3, lower, shows the length of sentences in *On the Origin of Species* (Darwin 1859). I want to report numerical values that will summarise these counts, and can achieve most of this with two sorts of measure: one is of the centrality of the data and the second of its dispersion (spread). Rarer measures are of asymmetry or lopsidedness (skewness,) and of squareness of the distribution (kurtosis). The two commonest measures of centrality are the (arithmetic) mean, $\mu = \Sigma x / n$, got by summation of the numerical values and division by the number of values, and the median, got by a sort of the numerical values into size order to reveal the central value $x_{(n+1)}/2$ if n is odd, or to allow the calculation of the arithmetic mean of the two central values $[(x_{(n/2)} + x_{(n/2)\ +\ 1}] / 2$ if n is even. The mean of the whole population (all the sentences in *On the Origin of Species*) is usually represented by Greek 'μ'; the mean of a sample from the population by '$\bar{x}$' ('x-bar'). There is no generally agreed symbol for the median, but $\dot{m}$ (m with a dot above – Unicode 1E41) is often used.

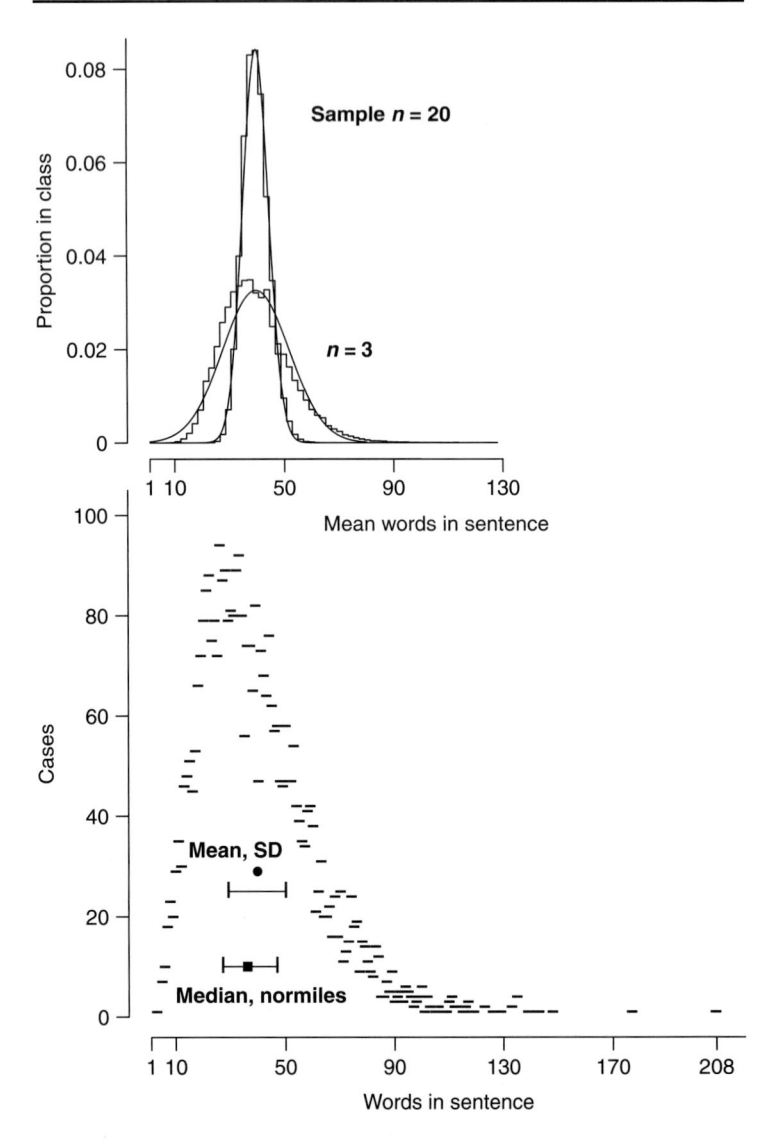

FIGURE 8.3 *Lower*: distribution of sentence length (number of words) in Darwin's *On the Origin of Species*. Mean, standard deviation (SD, a single-value range centred arbitrarily on the mean), the median, and the two normiles are shown. The normiles (see text) are specific values on the abscissa that define a range (NR) that, were the distribution 'normal', which it is clearly not, would be of the same length as the SD and would be centred on the median, which would be the same as the mean. (*cont.*)

If the distribution is symmetric and there are no wildly erratic outliers then about as many values will fall below the mean as fall above it, so the mean and median will be close to one another. But the calculation of the mean includes the specific numerical values of all the items, so if there are extreme outliers (or the whole distribution is strongly skewed) the mean value will be affected whereas the median will not. The median then gives a better idea of the 'centre' of the distribution, because it is much more robust (less affected) than the mean by moderate changes in the numerical values. (So why do we not use the median all the time to describe centrality?)

If you used the mean (and the distribution is near-'normal' ▼) then the usual measure of variation of a population, got from a *sample*, is the variance, s^2, defined as

$$s^2 = \Sigma(x-\bar{x})^2/(n-1). \tag{8.1}$$

▼ The 'normal' distribution is often called 'bell-shaped' but, unlike a bell, it extends to infinity in both directions. It is symmetric, but many distributions are symmetric and 'bell-shaped' yet their shoulders are broader or narrower than those of the 'normal'. The distribution was first discovered by de

Caption for Figure 8.3 (*cont.*) Upper: distribution (steps) of mean length of 100,000 *samples* of 3 and of 20 sentences taken at random, with replacement, from the population in the the lower part. The continuous lines are those for the same values of mean (of means) and SEM (standard deviation of means). For $n = 20$ the line is close to the samples; for $n = 3$ it deviates systematically. The inset Table below shows that for $n = 20$ the mean of sample means is the same (to three significant digits) as the population mean and that the SEM (the standard deviation of the sample means) is close to SD$\sqrt{n}$, even though the population is strongly skewed.

	Mean	SD	SD/$\sqrt{20}$
Population	39.6	21.6	4.65
Sample $n = 20$	39.6	4.74 (SEM)	

> Moivre in 1733, and closely approximates the distribution of many biological variables such as height and mass of individual organisms. For this reason it became known as the 'normal' distribution. Not until 1809 did Gauss apply it in his analysis of errors in geodesic observations, and this discovery lies at the root of many statistical techniques. I enclose 'normal' in single quotes to indicate this special meaning.

Variance is what statisticians work with, because it is mathematically tractable: variances can be added, and, conversely, can be partitioned (by subtraction) to reveal the effects of different influences. But as a measure of dispersion for descriptive use then $s = \sqrt{s^2}$, the standard deviation, often represented as 'SD', is more useful because it has the same units as the original measurements. On graphs one can then plot the SD as a distance along the same axis as the mean, though there is no necessity for the common practice of plotting SD as 'whiskers' centred on the mean. The SD is always positive, even if all the values on which it is based are negative.

The dimensionless 'coefficient of variation', $CV = s/\bar{x}$, perhaps expressed in '%', is often characteristic of a particular method and sampling situation. If the smallest possible numerical value of the variable is zero the CV may be useful for description or for tracking changes in time. In laboratories making routine analyses the CV of measurements made on Mondays or after other holidays is usually rather larger than it becomes after the routine is re-established. But variables such as temperature on the Celsius scale, and pH, have an arbitrary zero, so their CV is also arbitrary, which makes comparison with the CV of another variable meaningless. See Webster (2001) for other problems with CV.

In physics the CV of measurements is commonly $<$ 1e-4, and occasionally $<$ 1e-12. In bioscience, measurement CV is often 0.01 to 0.05, while sampling CV is often $>$ 0.2. The running CV may be useful in a laboratory making a large number of measurements by the same method, but it is rarely useful in publications.

If you use the median as the measure of centrality then the conventional measure of dispersion would be the quartiles: the *two* values of the variable that, with the median, divide the ordered observations as closely as possible into four equal groups of observations. Quartiles are arbitrary, and the distance between them is not directly comparable to the standard deviation. There is attraction therefore in choosing quantiles ▼ that give the same values for the distance between them as the standard deviation *if* the distribution is 'normal'. These normiles as we can call them are 0.50 ± 0.19 i.e. the 0.31 and 0.69 quantiles. These can be plotted on the same axis as the median, but note that while the SD is a single valued distance that can be plotted anywhere on the axis (Figure 8.3, lower), the normiles are two specific points on the axis. The distance between them ('normile range', NR) will be the same as the SD if the distribution is 'normal', but if it is not then the distance between them will usually be different from the SD, and the length of the parts on either side of the median may be different, reflecting the skewness. A simple measure of skewness k_m is '$(u - \dot{m}) / (\dot{m} - l)$', where u and l are the upper and lower quantiles. For a symmetrical distribution the value is 0.0. For the skewed *Origin of Species* data (Figure 8.3), the median and normiles are the more helpful measures: the NR is asymmetric about the median ($k_m = 1.2$), and is less than the SD, which is inflated by the tail of long sentences. (The argument is strictly for a continuous variable, whereas the number of words in a sentence is discrete. But there are so many discrete sentence lengths in *On the Origin of Species* that the data approximate those of a continuous variable.)

▼ 'Quantile' is the general name for a value (between 0 and 1) that includes a specific proportion of the observations. The 0.5 quantile is the median, and the 0.25 and 0.75 quantiles are the quartiles (with an 'r') that include half the observations between them.

(Im)precision and testing: standard error of the mean (SEM)

So much for description. Now turn to the imprecision of a mean value. The upper part of Figure 8.3 shows as stepped histograms the result of taking numerous random samples of 3 and of 20 sentences (with replacement) from the population of 3764 sentences in *On the Origin of Species*. The mean of the sample means is close to the mean of the whole population. The smooth curves are the 'normal' distributions with the same mean as the mean of the sample means and the same standard deviation as the standard deviation *of the sample means*. For $n = 20$ the 'normal' for the samples is close to the experimental distribution of sample means – remarkable given how skewed the population is – and is explicable by the Central Limit Theorem.▼ It is this standard deviation *of the sample means* that is called the standard error of the mean, $s_{\bar{x}}$ and measures the imprecision of the mean. In text it is often represented as 'SEM', or by contraction, as 'SE'. The similarity of 'SE' to 'SD' is one cause of the widespread confusion of the two, so I prefer 'SEM' (the context should prevent confusion with the scanning electron microscope).

▼ The Central Limit Theorem (CLT) states that the sum of a large number of independent random variables will be approximately 'normally' distributed, almost regardless of their individual distributions. This is why so many biological variables (affected by a large number of different and independent random errors) are close to 'normal'. It also underlies sampling, as is shown by experiment in Figure 8.3: the numerical values of the means of samples from a population usually follow a distribution close to the 'normal' *even if the parent distribution is strongly skewed*. The CLT was first discovered by Laplace in 1812, but his work was hard to follow and it was 1835 before Quételet showed how widespread was its application (Bulmer 1965).

For $n = 20$ the experimental distribution of sample means agrees closely with the calculated 'normal', but for the small sample $n = 3$ the distribution deviates systematically and the samples follow Student's 't' distribution (which includes n as a parameter and tends towards the 'normal' as n increases). This distribution was discovered experimentally by WS Gosset, who was employed by Guinness, the brewers. The company allowed him to publish his work but only under the assumed name: 'A Student' (Student 1908).

Large-scale experimental sampling by computer of the kind shown in Figure 8.3 is comforting but usually impractical in real life. Fortunately the standard error of the mean can be calculated simply. The variance of sample means is s^2 / n so the SEM, $s_{\bar{x}} = s/\sqrt{n}$. The table in the caption to Figure 8.3 compares the calculated and experimentally determined SEM. The SEM is a measure of the imprecision with which the mean is known, and is needed *whenever means are being compared* with each other or with some fixed numerical value (such as 0.0).

The fact that large-sample means follow the 'normal' distribution even though the distribution from which they were sampled is skewed is of supreme importance, and underlies a large part of conventional testing of hypotheses.

One can see from the definition of SD, s in [8.1], that on average the addition of another item inflates the sum of squares by the same average amount and increases n by 1, so the value of s remains about the same when more measurements are added: s is a characteristic of a particular sort of repeated measurement on a particular sort of object. But the value of the SEM diminishes as more items are added because s remains the same but the divisor, $\sqrt{n}$, increases. This reduction in SEM (decreasing imprecision) is what you get for your extra work as you increase the number, n, of measurements in the sample. As you increase n so you reduce your uncertainty about the value of the mean. But to decrease imprecision by a factor f you will need f^2 as many measurements. For example, to decrease imprecision 10-fold requires 100 times as many measurements.

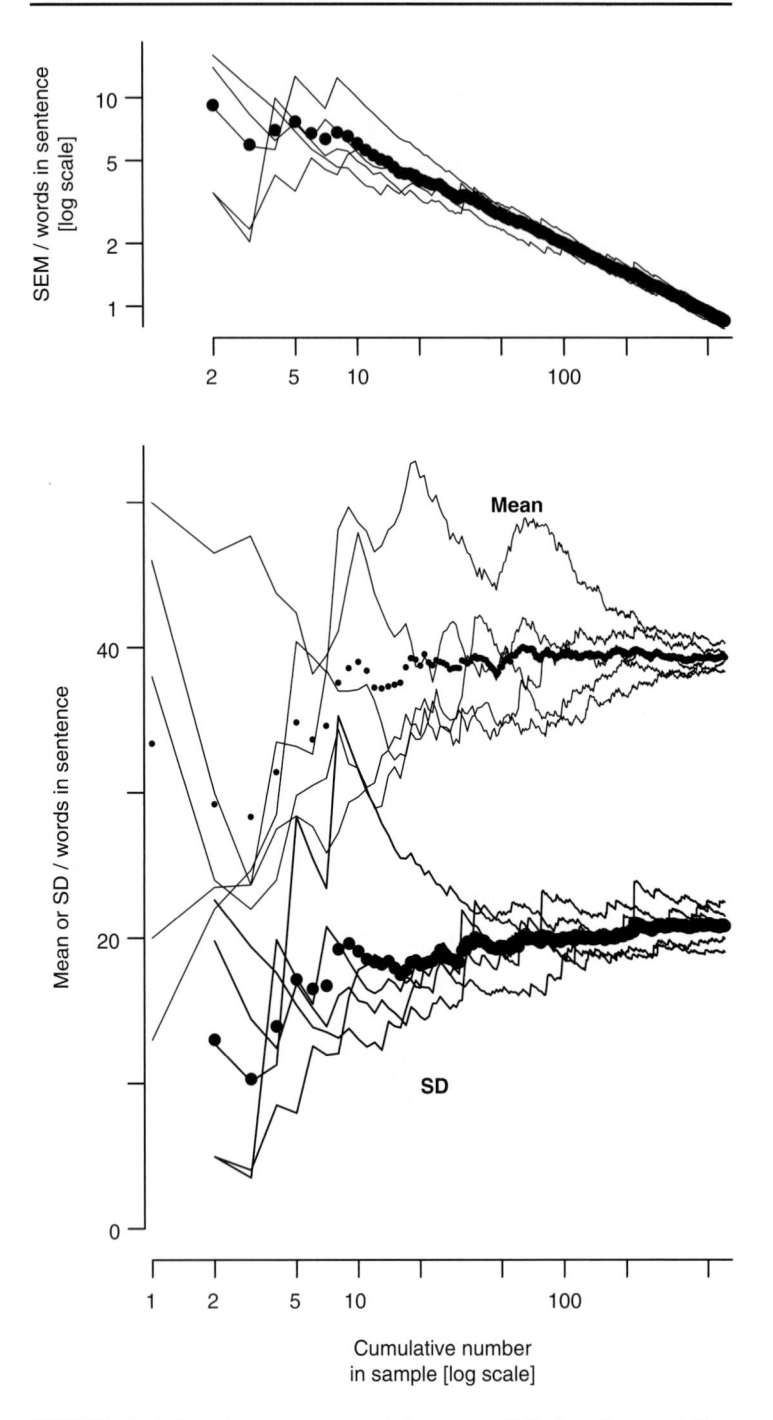

FIGURE 8.4 Sample sentences were taken sequentially from the population in the top half of Figure 8.3 (sentence length in *The Origin of Species*), *(cont.)*

The effects of taking two items, then adding a third, fourth and so on up to 500, and calculating mean $\bar{x}$, SD s, and $s_{\bar{x}}$ SEM after each addition, are shown in Figure 8.4. All three measures fluctuate erratically for small n, as one would expect. The fluctuations diminish as n increases because adding one more item when n is large can have only a small effect. For $n < 20$ the SD and SEM can be a long way from steady values. The SEM, standard error, graph is plotted on log–log scales so that one can see that SEM follows a slope of $-\frac{1}{2}$, as expected with a log scale and the $\sqrt{n}$ divisor.

The SD is often used for 'whiskers' about a point in a graph to show the dispersion of the measurements. But the SEM, also often used in the same way, is relevant only in the context of the question 'are two mean numerical value points significantly different?'. This practice is defensible if the two samples are large ($n > 30$, say). But for small samples such use is misleading because the distribution diverges from the 'normal', as we have seen for samples of $n = 3$ in Figure 8.3. What we need here is the confidence interval (CI) which is got by application of Student's t, which compensates the SEM for small n. For large n (say more than 30) then the 95% CI is close to the mean $\pm 2 \times$ SEM (Figure 8.5, right), because for 95% confidence that the true value lies within the CI the value of t is 1.96, which is close to 2. But the common practice of showing whiskers $2 \times$ SEM on either side of the mean for small n is misleading and the more so the smaller is n (Figure 8.5, centre and left).

Figure 8.5 shows, for three sample sizes, whiskers for range, SD, normile, SEM, and 95% CI. The type of whisker should *always* be

---

Caption for Figure 8.4 (*cont.*) and the mean, standard deviation, and standard error of the mean ($\bar{x}$, SD, SEM) were calculated for $n = (1), 2, 3, 4$ sentences, and so on up to 500. The process was repeated five times, results being shown by continuous thin lines. The mean of these five is shown by filled circles. The mean of means and standard deviations is little affected by n, but the mean of standard errors decreases as n increases. Note that for SEM both scales are log transformed and the slope is close to −0.5. Note also the precipitous leaps in the mean of small samples as one of the outliers is included.

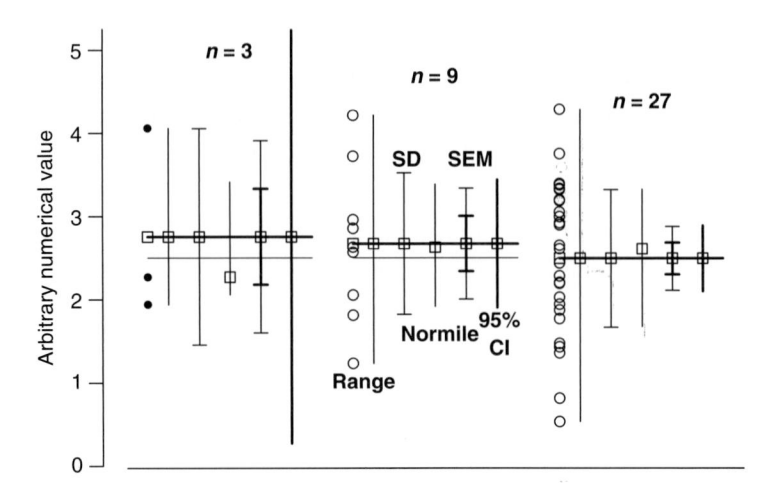

FIGURE 8.5 Common sorts of 'whiskers' applied to mean values (thick horizontal lines) of samples of three sizes (*left to right:* $n = 3, 9, 27$) drawn from 'normal' populations with mean 2.5 (thin horizontal lines) and variance 1.0. The values are shown by circles, filled for $n = 3$, but unfilled for the others so that the individual points can be distinguished. Vertical 'whiskers' from the left are: range, SD, lower and upper normile about the median (see Chapter 6, Measures of dispersion), 1 and 2 × SEM, and 95% confidence interval (CI). Open squares are at the mean or, for normiles, the median. Notice how, as n increases, the difference between sample mean and population diminishes; range increases; SD stabilises; normiles become less skewed and more nearly equal to SD; 95% CI decreases and becomes nearly equal to mean ± 2 × SE.

specified in the caption to a figure. The ways that these five change with sample size relative to one another is instructive.

Significant digits

Misunderstanding of the difference between descriptors and measures of imprecision, between SD and SEM, is common among bioscientists, but ignorance of significant digits is near universal.

A government website states, of a school with 600 pupils, "Percentage of all full-time equivalent teachers with qualified teacher status: 97.9497693490518". The implied precision is utterly ridiculous (a little thought also shows that it is impossible).

Just as implausible, when one considers the evidence it was based on, is the exactness of Bishop Ussher's calculation that the Earth was formed near nightfall preceding Sunday, October 23, 4004 BC. Even more egregious is the conclusion by four fourteenth-century mystics (names unrecorded) that the number of angels ▼ in heaven is exactly 301 655 722. It is not the size of this value that makes it ridiculous but the suspicion that the methods used cannot sustain the exactness implied – that the true value was ± 1. But the frequency of a transition of a trapped and laser-cooled, lone ion of $^{88}Sr^+$ is reported (Margolis *et al.* 2004), convincingly, as 444 779 044 095 484.6 Hz, with an SEM of 1.5 Hz. It is less obvious perhaps that the statement in a scientific article I have just read that a mean numerical value is 3.863 with an SEM of 2.162 (based on 17 measurements) must be treated almost as sceptically as the first three examples. The seven authors (and the referees and editor) of this short article have shown in the most conspicuous and long-lasting way that they are all innumerate.■ To see why their statement is so revealing we need to examine the relation between an SEM and the mean it is attached to.

▼ There is a large and conflicting literature about angels. They are generally supposed to have all come into existence on only one occasion, and to be immortal: a situation unfamiliar to population ecologists.

■ Aristotle understood the problem: 'It is the mark of a civilised man, and a hallmark of his culture, that he applies no more precision to a problem than its nature permits or its solution demands.' More recently R. V. Jones (1978), a pioneer of the use of science in military intelligence, wrote "... at the age of twelve ... a new physics master [set us too much homework]. I worked the answer out to thirteen places of decimals, knowing ... that this was unjustified. [When reprimanded by the master] I replied that I thought he would like an answer matching the length of the homework... The point of this story is that as fourth form schoolboys we already knew well how many places of decimals were justified in particular measurements: its significance was to be evident at [the bombing of] Coventry in 1940." (Passage reproduced from Jones RV 1978 *Most Secret War: British Scientific Intelligence 1939-1945*. London, Hamish Hamilton. 556 pp. Reproduced by permission of Penguin Books Ltd.)

Both Aristotle and the schoolboy Jones understood the need, though it would be a surprise if either knew the solution presented here.

Table 8.1(a) shows 60 values drawn at random from a population with mean 39.615 and SEM of 4.7 – the same numerical values as those shown in the upper part of Figure 8.3. In the '10s' decade all the digits are either '3' or '4' (the mean value lies high in the 30s close to 40). But in the '1's every digit from '0' to '9' appears between 1 and 12 times, and the true digit and its neighbours ('8', '9', and '0') appear in only 37% of instances – little better than the $18/60 = 30\%$ expected if digits in these positions were random.

In this instance we must conclude that only the first non-zero digit in the mean value (rounded to '4' for presentation) is significant, and that any others that follow it are essentially close to random digits and thus almost meaningless. The reported value would be 4*0*, the zero italicised to show that it is not significant but is there merely to show the position of the decimal point. To use italics for non-significant digits may seem a counsel of perfection. Some of the old literature used a smaller font size for the same purpose. I have never seen either convention followed in a modern bioscience journal. Here, however, is an example where italic for packing digits would be useful. The concentration (μg dm^{-3}) in $n = 7$ samples of dissolved inorganic nitrogen (DIN) had a mean and (SEM) of 57 and (9.2). Comparison was made with the concentration of dissolved phosphate of 45 and (7.7). Another comparison was

233 SIGNIFICANT DIGITS

TABLE 8.1 (a) 60 numerical values drawn at random from a 'normal' population with mean 39.615, SD 4.7 (simulating samples of the mean length in words of 20 sentences from Darwin's *On the Origin of Species*. The choice of 60 is arbitrary. The numerical values range from 28.0 to 49.5 (underlined) (b) Distribution of significant digits in a sample of 8000 values similar to those in (a) but simulating SEM 1.33. The target digit in each decade is underlined. The decades were followed until the digits were not significantly (χ^2, $P < 0.5$) non-random. The SigDig index is 1.4. '8000' is an arbitrary choice (c) As (b) but with SEM 0.0133, only 1/100 of that in (b). The SigDig index is 3.4. In the last row the target digit ('5', in 39.615) is not even the most abundant digit

(a)

38.7	32.2	36.3	38.8	32.8	37.3	43.3	42.0	38.4	38.5	32.6	36.3
28.0	38.4	38.9	39.8	44.6	35.2	39.6	42.5	42.0	39.6	44.6	37.3
37.2	43.6	46.3	36.1	42.9	33.5	35.7	40.4	48.1	35.9	38.5	47.9
38.2	46.0	33.8	49.5	34.8	39.4	43.5	39.1	36.6	35.1	38.8	39.9
39.7	35.9	42.4	42.3	36.5	42.4	36.2	41.2	43.8	42.4	49.0	38.9

(b)

	Digit									
Decade	0	1	2	3	4	5	6	7	8	9
10s	...	...	...	4846	3154	...	...	...	...	...
1s	1963	890	261	36	5	33	190	677	1621	2324
0.1s	827	798	787	775	769	777	859	810	840	758

(c)

	Digit									
Decade	0	1	2	3	4	5	6	7	8	9
10s	...	...	...	8000	...	...	...	...	...	...
1s	...	...	...	...	...	...	...	...	...	8000
0.1s	...	...	...	...	...	924	7076	...	...	...
0.01s	1738	2331	1825	894	240	43	5	35	164	725
0.0001s	759	823	787	806	800	768	797	827	813	820

made with the concentration of dissolved total nitrogen reported as 11,958 and (923). The precision implied by these last values is ridiculous. One solution would have been to change units and report 12 and (9) mg dm^{-3}. But this makes comparison with DIN less easy. Better would have been to report the concentration of dissolved total nitrogen as 12*000* and (9*00*) μg dm^{-3}.

This experiment is extended in Table 8.1(b) which shows 8000 samples, binned by digit in the 10s, 1s, and 0.1s from a population mean of 39.615 and SEM of 1.33. The digit in the 10s decade is clearly meaningful, and so is that in the 1s, but in the 0.1s decade, the target digit '6', though the most frequent in its decade, is barely better than random: a mean of '39' is worth reporting, but '39.6' is overdoing things.

Table 8.1(c) derives from the same mean but an SEM 100 times smaller at 0.0133. This supports 39.61, but in the next '0.00x' decade the target digit, '5', is not even the most frequent in its decade. In Table 8.1(a) the counts in the 0.1s decade are near random, but if we decrease the SEM gradually the totals for each digit in a decade become more and more unequal as peaks emerge and grow from the hummocky, slowly sinking, plain of other digits and, consequently, indicate that we may soon be able to justify another significant digit. In a report, the number of significant digits must be integer, but to understand the trends we need an index, D_M, that is at least semi-continuous. This index is derived in Clymo (2012), where the 'rules' in Table 8.2 summarising significant digits emerge. The simple rule is that the mean should be stopped at the decade in which the first significant (non-zero) digit of the SEM is found. The same article derives the simple rule, also in Table 8.2, for significant digits in the SEM itself.

What can you do if you have only one measurement? You can assume a SEM that is about what the SD might have been. For biological material and measurements then about 25% of the value is often a reasonable guess to start with. A measured value of 53.77 would suggest a SEM of 11. The measured value would then be reported with one significant digit as 5*0*, the italic '*0*' being of no significance except to show where the decimal point is.

These rules apply when the discrimination – the number of digits in individual measurements – is more than that given by the rule. But if, for example, you have measurements recorded to the nearest centimetre then neither the SEM or mean justifies more precision than a whole number of centimetres.

SIGNIFICANT DIGITS

TABLE 8.2 Rules determining the number of significant digits to report. See Clymo (2012) for justification

Rule 1: For significant digits (D_M) in the mean
The *last* significant digit in the mean is in the same decade as the *first* significant digit in the SEM; but *Rule 1 extra (you may ignore this supplement without serious error)*:
If the first significant digit in C = mean/SEM is '3' to '9' then one more digit is significant in the mean.

Rule 2: For significant digits (D_{SEM}) in the SEM itself

n in sample	2 to 6	7 to 100	101 to 10 000	10 001 to 1e6	> 1e6
Significant digits, D_{SEM}	1	2	3	4	5

Rule 3: For counts as percentages
For fewer than 100 observations the two digits in a percentage overstate the precision; for more than 100 (assuming counting statistics) *Rule 1* applies.

n in sample[a]	11 to 20	21 to 50	51 to 100	101 to 10 000	10 001 to 1e6
Report % to the nearest /%	5	2	1	0.1	0.01

[a] For fewer than 10 observations do not use %
Examples: 7 / 17 = 40% (not 41.17%); 6 / 17 = 35%

TABLE 8.3 Deciding the number of significant digits

Raw mean	24.18496	24.18496	24.18496	24.18496
Raw SEM	0.83662	0.43662	0.0008401	87.2295
Mean / SEM	28.8	55.4	28790	0.277
n	15	23	5	148
Reported SEM	0.84	0.44	0.0008	87.2
Reported mean	24	24.2	24.185	2*0*[a]

[a] The italic '*0*' in the '1s' decade of the last entry in the table is a non-significant packing zero which shows where the decimal point would be. Also, by convention, the mean always has at least one non-zero digit, but if this digit is not significant then it too can usefully be shown in italic.

Table 8.3 shows a few examples of the application of the rules.

Here is a practical example. An ecologist interested in the rate at which organic matter was added to woodland soil set out 30 litter

TABLE 8.4 Mass ± standard error (g m^{-2} yr^{-1}), $n = 30$, of fractions caught in litter traps each 50 cm × 50 cm. Versions 1 to 4 show the SEM in four different formats; version 5 shows 95% CI as sub- and superscripts

	As published	Version 2	Version 3	Version 4	Version 5
Leaves	15.196 ± 0.462	15.2 ± 0.46	15.2 (0.46)	15.2$_{0.46}$	15.2$_{14.4}^{16.0}$
Twigs	1.600 ± 0.288	1.6 ± 0.29	1.6 (0.29)	1.6$_{0.29}$	1.6$_{1.01}^{2.19}$
Flowers	0.004 ± 0.002	0.004 ± 0.0015	0.004 (0.0015)	0.004$_{0.0015}$	0.004$_{0.001}^{0.007}$
Total	16.8	16.8	16.8	16.8	16.8

traps – shallow boxes to collect flowers, leaves, and fallen twigs and small branches – and recovered the trapped material after three months. He dried the contents of each box and weighed the total, then removed twigs and weighed them, then removed flowers and weighed them. Leaves were calculated by subtraction of the mass of twigs and of flowers from the total mass. The published results are in Table 8.4. The '00' at the end of the 1.6 for twigs is entirely cosmetic, added by a copy-editor, so that all the numerical values have the same number of digits after the decimal point, and with no justification in the measurements. The '96' at the end of leaves was what the author submitted, calculated to make the sum correct to three places after the decimal point. Three improved versions follow in Table 8.4. Version 2 reports significant digits only but looks less neat. Version 3 improves neatness by aligning the means on the decimal point. It also illustrates the use of parentheses instead of '±'. Version 4 places the SEM as a subscript, aligns on the decimal point, and takes less space. The fifth (last) version shows the mean and 95% CI, which can be calculated from the information given. But why put readers to that trouble?

During calculations carry as many digits as necessary to avoid cumulative rounding errors, but the precision with which a value is known (or guessed) should always be used to determine the digits to be reported even though this can result in a column of numerical values in a table with different numbers of digits after the decimal point. Resist editorial attempts to enforce uniformity where the precision does not justify it. ▼ It is *your* reputation that is at risk.

> ▼ You might adopt, or at least adapt, the stirring advice in the Church of England late-evening office of Compline: '... brethren be sober, be vigilant, for thine adversary the devil goeth about as a roaring lion, seeking whom he may devour, whom resist, steadfast in the faith.'

Combining errors

In Table 8.4 I slid over a problem. The SEM of the total mass and the SEM of the flower mass are both known. But we have no direct measurement of the SEM (imprecision) of leaf mass because it was calculated by subtracting two of the three components from the total. So how, in general, can one assess imprecision in a variable that is calculated rather than measured? Of course one must have calculated the mean and SEM for all the measured variables. The general idea is shown in Figure 8.6

The variable z depends on two other variables u and v. At the particular point shown in the centre of the graph the surface slopes steeply along the u-axis. A small change in u will therefore produce a large change in z. But at the same point in the v direction the surface slopes only gently so a similar proportional change in v will have little effect on z. It seems intuitive that the overall error should be more affected by u than by v. The formalities may not be known or easily available to bioscientists, so I give the details here.

For 'normal' distributions, which we have seen that sampling commonly generates, then a surprisingly simple and critically important fact is that if you add two distributions then the variance of the combined distribution is simply the sum of the individual variances. The variance required here is $(SEM)^2$. To calculate the variance in z one therefore takes the known variance of u and weights it by the (large) slope along the u direction at the

238 MANAGING ERROR

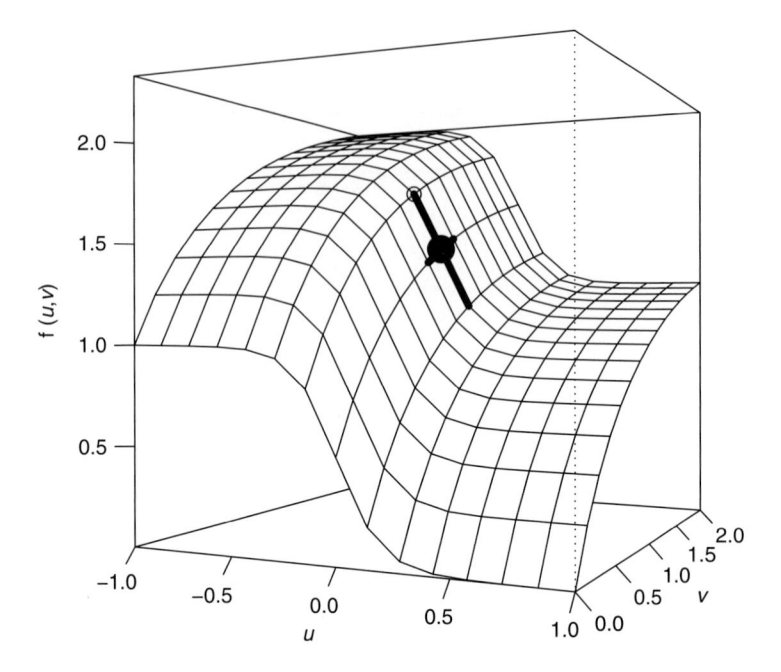

FIGURE 8.6 Dependence of z on u and v. At the central point the slope in relation to u is steep but in relation to v is shallow. The equation is z = 1 / (1 + exp (8 u)) + 2 v / (1 + v).

point, then add the variance in v weighted by the (much smaller) slope along the v direction. The slope is the partial differential coefficient (∂v), where v is the variable, got by differentiating the defining equation assuming everything but one variable is constant.

Formally, if $z = f(u, v)$ then

$$dz = \frac{\partial f}{\partial u} du + \frac{\partial f}{\partial v} dv. \tag{8.2}$$

Square both sides to give

$$dz^2 = \left(\frac{\partial f}{\partial u} du\right)^2 + \left(\frac{\partial f}{\partial v} dv\right)^2 + 2 \frac{\partial f}{\partial u} \frac{\partial f}{\partial v} du.dv. \tag{8.3}$$

The small change dz can be replaced by the SEM ($s_{\bar{z}}$), and du and dv can be replaced by $s_{\bar{u}}$ and $s_{\bar{v}}$ for imprecision, or a similar measure for overall accuracy including bias. 'Small' means that $s_{\bar{u}} / u$ and $s_{\bar{v}} / v$ are 0.05 or smaller. If the errors in u and v are uncorrelated (independent) and symmetric (equally likely to be positive and negative) then the last term, involving $du.dv$, will be close to zero, because while the first two terms are squared, and are therefore always positive, the third term is sometimes positive, sometimes negative, summing to near zero.

$$s^2_{(z)} \approx \left(\frac{\partial f}{\partial u} s_u\right)^2 + \left(\frac{\partial f}{\partial v} s_v\right)^2. \qquad [8.4]$$

This says that to get the square of the error in z you weight each component error by its partial differential coefficient, square the result, then add the squared values together. As the individual terms are squared, so the errors are always positive, i.e. the total error always increases as items are added even if the item in the original equation is subtracted. An example of this is the calculation of overall error in the count of a sample whose radioactivity only slightly exceeds background. The mean value is got by subtraction of background count rate from sample rate, but the overall error is got by addition of the components from sample and background. (It will probably be obvious that in these circumstances we want the SEM of the background to be near that of the sample, so one should spend as much time counting background as counting sample.)

The same equations can find the proportion of time to be allotted to counting background and sample to achieve the smallest possible imprecision with a given total counting time. Nowadays these calculations are usually performed automatically by the equipment: the days of personal struggle with the algebra and a slide rule or other calculator are gone though the principles remain the same.

This approach can be extended to three or more variables. If $z = f(u, v, w, ?)$ then

$$s^2_{(z)} \approx \left(\frac{\partial f}{\partial u} s_u\right)^2 + \left(\frac{\partial f}{\partial v} s_v\right)^2 + \left(\frac{\partial f}{\partial w} s_w\right)^2 + \ldots \qquad [8.5]$$

Examples

The following example shows how these equations can be used in planning. I needed to assess the volume of gas space in peat. The equation for this, derived from the fact that $V_t = V_p + V_w + V_g$, is:

$$P_g = \frac{V_g}{V_t} = 1 - \frac{m_p}{\rho_p V_t} - \frac{m_t}{\rho_w V_t} + \frac{m_p}{\rho_w V_t} \quad [8.6]$$

where V is volume, m is mass, ρ is intrinsic density of the solids in peat, and the subscripts are 'g' for gas, 'p' for peat dry substance, 'w' for water, and 't' for total. P_g is the proportion that gas forms of the whole – the required variable.

This equation contains five variables (V_t, m_t, m_p, ρ_p, and ρ_w) that can be measured and for each of which the SEM can be measured, or guesstimated with sufficient accuracy for planning. The partial differential coefficients for all five are easily derived ▼ after which typical values for the measurements and their SEM can be inserted, as in Table 8.5, which constitutes a sensitivity analysis.

> ▼ For $y = x^n$ then $\partial y / \partial x = n\, x^{(n-1)}$. This works for negative n too and is all that is needed for the example here.

The last column shows the unsquared contribution of the five measured variables. The overall error – the square root of the sum of squares of these entries – is 3.1%, which was too large for the intended purpose. Inspection of column 8 shows that much the largest contribution comes from V_t, the total volume. This is determined by how carefully the equipment is operated in the field. Having thus been alerted, I managed to reduce the SEM of V_t to about 2% in practice. The next biggest contribution comes from m_t, the total mass. In practice this measurement, using a standard open-pan balance, had an SEM smaller than that assumed in planning. The 5% SEM for ρ_p, the intrinsic density of dry peat, which I assessed by a separate method and which had seemed worryingly large, turned out to be of little importance – in this work at least.

EXAMPLES

TABLE 8.5 Sensitivity analysis of a planned method to measure the proportion of gas in peat

Column 1 = C1	C2	C3	C4	C5	C6	C7	C8[a]
Variable	Symbol	Typical value	Partial differential (PD) $\partial P_g / \partial ?$	Value of PD	SEM	SEM (%)	C5 × C7[a]
Total volume	V_t	2050 cm^3	$[m_p/\rho_p+(m_t-m_p)/\rho_w]/V_t^2$	4.64e-4	3	61.9	0.029
Total mass	m_t	1991 g	$-1/\rho_w V_t$	-4.89e-4	1	19.9	-0.010
Dry peat mass	m_p	123 g	$1 / [V_t (\rho_w-\rho_p)]$	8.11e-4	1	1.23	0.001
Density[b] of dry peat	ρ_p	1.60 g cm^{-3}	$m_p / (\rho_p^2 V_t)$	2.34e-2	5	0.08	0.002
Density of water, 20 °C	ρ_w	0.998203 g cm^{-3}	$[(m_t-m_p)/V_t]/\rho_w^2$	9.14e-1	0.001	1e-5	0.000

[a] Column 8 is the (unsquared) contribution to overall imprecision.
[b] Intrinsic density.

A second example of the combination of errors may serve to convince you of the value of this approach. Governments increasingly try to assess the total load on their countries of some 'pollutant': let us suppose the metal nickel. The concentration of the general influx in rain and dust is tiny but suppose that there are two smelters – point sources – that cause influx of high concentrations though over an area that is only 1/100 000 of the total country. Chemical methods can measure the high concentrations easily, with an SEM of only 1% of the mean value or smaller. But the 10 000 times more dilute background influxes tax the analytical chemist's skills and have an SEM that is 10% of the mean. Which of these two analyses – the point source or the background – will dominate the estimate of overall influx? The low-concentration source is much the most important (as you can see by making a sensitivity analysis similar to, but simpler than, the previous example), and efforts to improve the overall estimate should concentrate on improving the analytical method for low concentrations.

9

Data interrelations

Interpreting *P* values

Suppose you have just finished the measurements in a simple experiment to compare the abundance of weeds on 1-m^2 plots sprayed either with a weedkiller in solution or with the same volume of water alone. You adopt the null hypothesis: that the weedkiller has no effect, and make a suitable test of this hypothesis (*t*-test or *U*-test, as appropriate) obtaining the result $P = 0.04$. That is to say that, if the hypothesis is true, then in a large number of similar trials one would expect to have got mean values at least as different as these were in about 4% of trials. The statistics stops at this point. So what next?

Anything further is a judgement by you and your readers. A convention has grown that for routine work of no special importance, if $P \leq 0.05$ then, as practical people we think we will make fastest progress by concluding provisionally that the hypothesis is false while accepting that on average 1 in 20 of these provisional conclusions will be incorrect. But the value 0.05 is arbitrary.

- If we had only four digits on each hand it seems likely that the value would be $3 / 8^2$ (= 0.047 in base 10) rather than $5 / 10^2 = 0.050$.
- If the experiments are important and cannot be repeated for some reason (expense, no more material, unavailable equipment) then

you may argue that you are going to reject the hypothesis provisionally with P as high as 0.1 (for example).
- If your life depends on avoiding *wrongly* rejecting the hypothesis then you might be inclined to require that P be $\ll 0.000\,000\,1$.
- And if your life depends on avoiding wrongly rejecting the hypothesis yet there is a reward of 1 000 000 €/£/$ for correctly rejecting it, then you might be inclined to gamble at $P < 0.000\,1$ or even $P < 0.001$. This is the sort of subconscious judgement you make every time you get into a car: the weighing of advantages and risks of an adverse outcome.

The '$P = 0.05$ for routine work' is a ritual, an unspoken agreement: get a value smaller than that in uncontroversial work and your referees, editor, and readers will agree to your rejecting your hypothesis without serious argument. But do not tie yourself to this totem pole if the circumstances require more freedom.

If there is insufficient evidence to reject the hypothesis, does that mean it must be true? Certainly not! Absence of evidence is not evidence of absence. If you have measured several hundred instances of two treatments you may want to argue that at least it seems plausible that your hypothesis may be close to the truth. But if you measured only, say, four instances and the measurements show large variation then all you can conclude is that you made too few measurements to be able to reach *any* defensible conclusion. This sort of mistake is sometimes called a 'Type 2' (or 'Type II') error.

Type 1 error = α (Greek alpha) = probability of *wrongly rejecting* the null hypothesis, similar to convicting an accused person, initially assumed innocent, who actually *is* innocent;

Type 2 error = β (Greek beta) = probability of wrongly NOT *rejecting* the null hypothesis, similar to acquitting an accused person, initially assumed innocent, who actually *is* guilty.

The reported value of P after a test of significance usually implies α alone, but when planning (at least) the 'cost' of wrongly *not* rejecting the null hypothesis can seem as great or greater than that of wrongly rejecting it: in our planning we wish to steer between Scylla and Charybdis, both the sea monster and the whirlpool. How can we do this? Rather easily for large n. 'Statistical power' is $(\lambda \alpha \sqrt{n})/\sigma = 1 - \beta$, where λ is the difference between the numerical value of the mean of the treated and untreated, n is the number of such differences, and σ is the population standard deviation of the differences. It may help to think of λ / σ as the treatment effect expressed in standard deviations. Define $Q = \alpha / \beta$, that is the quotient of the relative importance of Type 1 and Type 2 errors. Then the equation becomes

$$(\lambda / \sigma) \sqrt{n} = (1 / \alpha) - Q. \qquad [9.1]$$

You choose α (conventionally 0.05 for routine work) and decide Q (1.0 if Type 2 errors are of the same importance to you as Type 1 ones; 0.2 if you think they are five times as important, 5 for five times less important). Then, with two of λ, σ, and n known or assumed the third can be calculated. Most commonly one needs to get n to decide the scale of the experiment, but sometimes one may wish to know how big the treatment difference will be for a given n and σ.

Most apprentice scientists rapidly discover that the more variable the material is for a given difference in mean values the larger is the number of measurements needed to reject the hypothesis (to be able to claim that the experimental treatment has an effect) at a given value of P. To halve the SEM one needs four times the number of measurements; in general to improve precision by a factor f then one needs f^2 times as many measurements. Of course you must remember that if you make a large enough (10 000?) number of measurements you can expect to reject the hypothesis of 'no difference' for a tiny actual difference. But that does not mean that the difference is of any scientific importance.

A decision to reject carries with it the recognition that if you make a large number of such decisions with $P \leq 0.05$ then about 5% of the time you will wrongly reject the hypothesis and someone will have to go back and reinvestigate when the error begins to reveal itself. That specifies the compromise of practical people: a few mistakes but right most of the time. But be careful to avoid the consequences of a statistical fishing expedition. Suppose you have made measurements on $n = 46$ variables at several tens of sites, and then calculate the $n\,(n-1)\,/\,2 = 1035$ linear correlations between the variables, seeking to discover which are strongly correlated either positively or negatively, and willing to accept a high correlation wherever it appears. If you were to simulate the measurements with a set of uniformly distributed random numbers then among these 1035 correlation coefficients about 50 would have $P \leq 0.05$ – even though there are no meaningful correlations at all. Because an ounce of practice may seem better than a ton of theory, a similar example is shown in Figure 9.1. There you see that 10 000 ranked values of P derived from 2500 analyses of variance on random data are close to, but not exactly on, a straight line, and 492 of these (almost 0.05 of the 10 000) have '$P \leq 0.05$'.

This is the subject of 'mass significance'. As a rule of thumb you need to reduce the P value in proportion to the number of tests you make. So in the example above you may reject the hypothesis of no correlation for instances with $P \leq 0.05\,/\,1035 = 0.00005$. The precautions needed when fishing are much better known than they once were, and statistical programs now include 'post hoc', 'honest', 'Tukey' or 'Bonferroni' tests of significance which allow for this effect.

Are two means significantly different?

This is perhaps the simplest example where a hypothesis test is used, yet mistakes in handling it are surprisingly common. If data comprise two columns of values then t-testing is widely employed. A computer will make the calculations more quickly than an accident can happen, yet the results can be just as bad. Consider some of the possibilities.

1. The usual hypothesis is that there is no difference between the means, and that any difference may be positive or negative (a two-sided test). But others are possible. One is that the difference has some specific non-zero value (positive or negative). Another is

ARE TWO MEANS SIGNIFICANTLY DIFFERENT?

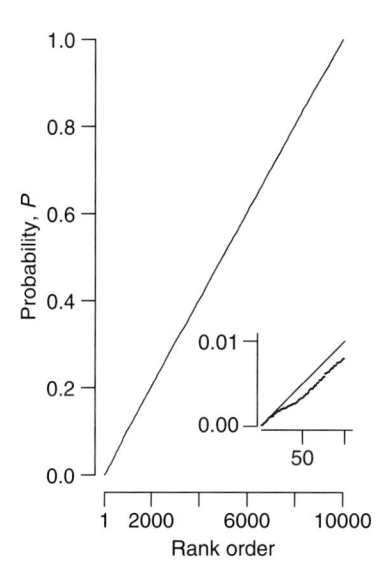

FIGURE 9.1 An analysis of variance was made of an imaginary experiment with three factors and two replicates:

AoV skeleton						
Factor	Levels	Effects	Deg F	Mean Sq	F	P
A	2	A	1	~	~	~
B	2	B	1	~	~	~
C	3	C	2	~	~	~
		AB	1	~	~	~
		Error	18	~		
		Total	23	~		

Main effects and the AB interaction (only) were included. This analysis was provided with data consisting of 24 random numbers distributed uniformly between 0 and 1. The mean square (MS), F and P values were calculated in the usual way, the analysis yielded four P values, shown by '~', which were saved. Another set of random data was taken and the analysis repeated, and so on 2500 times. The 10 000 values of P were then ranked and are shown plotted against their rank order. Inset are the lowest ranked 100 with the axes expanded with a reference line of slope 1.0.

that the difference is simply positive or that it is simply negative (one-sided tests).
2. The t-statistic is calculated as '(Difference in means)/SEM', but what difference and what SEM depends on the circumstances. You can recognise two general situations.
    a. Small numbers (say 5 to 20) of *paired* observations. This is the most powerful situation in that it gives the smallest value of P for a given set of values, and hence the greatest opportunity to reject the null hypothesis. The pairing will have been part of the experimental design. In the comparison of weedkiller with the same volume of water you may have arranged pairs of adjacent plots, likely to have very similar soils and vegetation, one given weedkiller and the other water. Such pairs of plots were distributed at least 100 m apart in a large field thus sampling a variety of differences in soil and vegetation. In medical research one may match pairs of people for characteristics such as age, gender, weight, or perhaps make measurements on, for example, reaction time of the same individual before and after some treatment.

    You calculate the single column of n differences, Δ, between pairs. The commonest hypothesis is that the mean difference is zero. Then t is the mean Δ divided by the SEM of the differences.
    b. Small numbers (say 5 to 20) of *unpaired* observations. Weedkiller- and water-treated plots were scattered over the field randomly in space and independent of one another, so you cannot match a particular weedkiller-treated plot with a particular water-only one. There may be different numbers of the two treatments (different number of values in the two columns). Let the weedkiller plots be u and the water ones be v. You calculate the mean u and the mean v, then the appropriate SEM. If the variance of u and v is similar the SEM calculation is simple. If the variances are markedly different the calculation is a bit more complicated. (A computer program may make the appropriate calculation automatically.)
3. Sometimes the values in one or both columns are strongly non-'normal' (perhaps skewed or with outliers). Here the underlying conditions, on which t tests are based, are not satisfied.

249 ARE TWO MEANS SIGNIFICANTLY DIFFERENT?

TABLE 9.1 Grasses /% on east-facing and west-facing slopes in 11 different places (data in rows rather than columns to save space). Two pairs, marked '*' and '**' are moderate and extreme outliers

| East-facing | 60 | 45 | 56 | 48 | 59 | 81** | 75* | 39 | 61 | 53 | 52 |
| West-facing | 75 | 46 | 71 | 52 | 57 | 93** | 70* | 61 | 67 | 60 | 59 |

TABLE 9.2 Probability, using the same data (see text) and *t*-test (Welch) or non-parametric (N-P) test (Wilcoxon–Mann–Whitney). The tests were two-sided

	t-test	N-P test
Paired	0.007	0.014
Unpaired	0.137	0.122

a. It may be possible to transform a positively skewed set of data – by taking the square root or, more extreme, the logarithm of the data values – so that they become satisfactorily 'normal'. See Webster (2001) for details.

b. If no transform is satisfactory you need a non-parametric test i.e. one that does not assume a distribution (*t*) with parameters. The calculations involve rank orders and differences.

In Table 9.1 are the percentage of grasses on east-facing and west-facing slopes in 11 different sites. The design allows paired testing, and as there is no clear *a priori* reason for expecting the percentage to be larger on east-facing than on west-facing slopes, or the reverse, so a two-tailed test is needed. For illustration I show results in Table 9.2, as *P* values for paired and unpaired, *t* and non-parametric tests. The extra power of the paired tests is clear: always try to use pairs if possible. If the data were 'normal' the parametric *t*-test would be more powerful in rejecting the hypothesis (H : 0) that there is no difference between slopes (it would produce a smaller *P* value than a non-parametric test), but here 2 / 11 of the slopes are outliers in the data (Table 9.1) so the non-parametric test produces a slightly smaller value of *P*. Of course, in a particular situation only one of these four tests would be chosen – the outliers invalidate a *t*-test – and the paired non-parametric one is the one to use.

The same principles apply when comparing statistics other than the mean.

QUESTIONS:
- Have you used a test of a two-treatment null hypothesis recently?
- Which of the four possibilities did you choose?
- Was that the right choice?

Correlation, regression, and functional analysis

Because the calculations for correlation and regression are so similar they are often confused. Yet they have different purposes, are often used unsuitably, and may both be unsuitable. Consider two columns of data with the same number of rows, as in the hypothesis testing in the previous section.

Correlation

Correlation is needed when you ask 'Does U increase as V increases, or U increase as V decreases?'. Neither U nor V is thought of as the 'cause' of the other (though it may be so in fact).

There are well-known techniques for calculating the strength of a correlation (a) when you assume that the distribution of both U and V is 'normal' and the relationship is linear (or at least that it is monotonic i.e. that points rank ordered on one axis will be rank ordered, in increasing or decreasing rank order, on the other as in Figure 9.2, lower right), and (b) if you make no assumptions about the distribution, but still assume a monotonic relation. These are respectively the Pearson product moment, and the Spearman non-parametric methods.

If you can see in the data that the relation goes up in one part of the range but down in another (Figure 9.2, upper left) then there is

251 CORRELATION, REGRESSION, AND FUNCTIONAL ANALYSIS

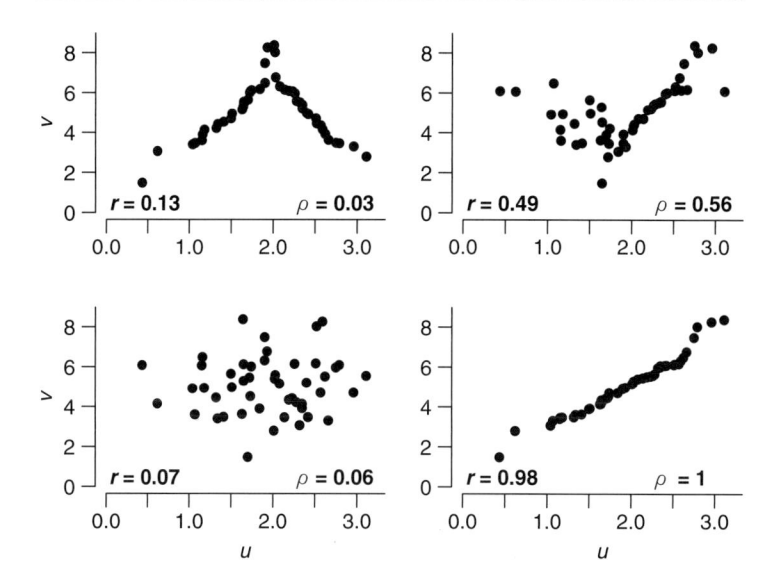

FIGURE 9.2 Two sets ('u' and 'v') of 48 normally distributed values were drawn at random with SD 0.6. The means were 2.0 and 5.0. These data are thus known to be suitable for correlation calculations. *Lower left*: as expected the correlation (r the Pearson coefficient, ρ the Spearman coefficient) is close to zero. The same data were then re-associated in various ways. *Lower right*: both sets were ranked. *Upper left*: they form a peak or caret. *Upper right*: some points form a clear linear relation and the rest form a cloud to their left. Correlation is not a sensible choice for the two upper arrangements as correlations near zero obscure very obvious non-linear patterns. Always examine data patterns before making analyses.

no point in linear correlation. Always examine data visually for obvious patterns before making further analyses.

Regression

Use regression when you want to *predict* one thing from a measurement of another. The predicted (dependent) variable is plotted on the ordinate (usually the vertical axis of a graph). For example, measuring the rate of photosynthesis, A, in a sample of water from a lake is a lengthy and technically complicated business, and measuring it at several depths to obtain a total A_T for the water column is even more difficult. A limnologist discovered that in carefully defined circumstances there was a close curvilinear relation between A_T and the depth at which a 30-cm diameter white disk (a Secchi disk) was still just visible as it was lowered into the lake. Using the regression of Secchi-disk depth to predict the rate of photosynthesis in the water column then became a much quicker and cheaper surrogate for the more difficult method.

If you use the simple methods, you assume a 'normal' distribution (often but not always true) and a linear relation, but there are also methods for curvilinear regression. You assume that all the uncertainty is in what is predicted, and minimise the sum of squares of distances in the vertical direction from each data point to the regression line. The predictor may itself have a 'normal' distribution (as in the Secchi disk example), or it may be assumed to be known exactly (as when you have chosen the temperature to apply in an experiment).

Total variance of the dependent variable (call it s_y^2) uses the differences from the mean. A similar and almost always smaller quantity (call it $s_{y.x}^2$) can be calculated using differences from the regression line. It is a surprising fact that $(s_y^2 - s_{y.x}^2) / s_y^2 = R^2$: the proportion of the total variance accounted for by the regression line is numerically the same as the square of the Pearson correlation coefficient. If you are using regression you have already assumed that there is some relation between the variables and will rarely need to quote the Pearson correlation coefficient, r, but the numerical square of that coefficient is a useful 'figure of merit'. This is the cause of a lot of confusion, for which reason it is often represented, as here, by uppercase R^2.

Notice the strange position of a chemical analyst's calibration line. The standards used are assumed to be exact, and all the error is in the readings. So you predict the readings from the standards. But then what you want to do is to use a reading from an unknown sample to 'back-predict' what standard would have given that reading. There is a solution to this problem (Webster 1997), but it is not the one usually used by chemical and physical analysts who often simply want a number without worrying much about its uncertainty because the analytical method is so much more precise than variation in samples to which it is applied. Bioscientists, however, often face larger and non-negligible variation in their calibration lines.

It is not necessary for the predictor to be in the physical sense the 'cause' of the dependent variable (though it often is). There may be some underlying cause of both the predictor and the predicted, for example 'intelligence test' score and shoe size in young humans (both related to age). This distinguishes regression from functional analysis (function fitting) where specified, possibly complex, cause is the essential feature. Many authors wrongly refer to regression when their purpose is really functional analysis.

Functional relations and analysis

Function fitting is needed if you want to understand the causal relation between two or more variables. For example, the relation between the rate of decay and temperature above the freezing point of water is often close to exponential. We are interested in the parameters (coefficients) of this relationship for what they can tell us about the process of decay. In linear regression you are interested in the intercept and slope only as means to predicting the dependent value; in a straight line functional relation you intend the values of intercept and slope to tell you things about the process. In the simplest case of a linear function and the abscissa values known exactly the calculation is the same as regression, but more usually it is not. Seek advice from Sprent (1969) and probably from a practising professional statistician.

Webster (1997) gives an excellent account of these things. His examples are from soil science but his account is completely general (and solves the analyst's problem described above).

To illustrate these differences consider Figure 9.3 which shows the annual mean flow of the Paraná river in South America and the annual sunspot number.

Rather surprisingly, at first sight, these two variables do seem to be related. The Pearson $r = 0.78$ with a probability $P \ll 0.001$ that

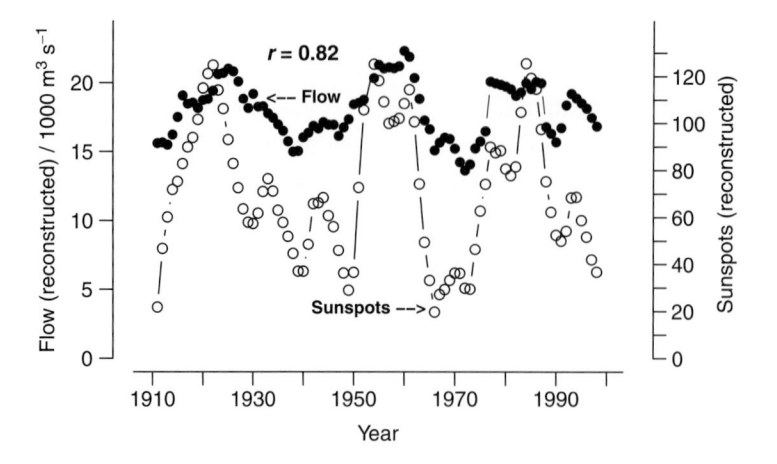

FIGURE 9.3 Yearly mean flow f (filled circles) of the Paraná river, and sunspot activity s (unfilled circles) from 1910. I have reconstructed the original values *approximately* from the standardised published data (Mauas, Flamenco & Buccino 2008). P for the Pearson correlation coefficient was $\ll 0.001$.

this is zero confirms the strong relation. For two years the river flow is missing but the sunspot number is known so we could now use the linear regression (Figure 9.4) of flow on sunspot activity to predict the flow for years for which the data are missing. After that we begin to wonder if there is a causal relation. We discover that there is a known relation between sunspot activity and solar radiance, and between solar radiance and rainfall. Eventually we may (I speculate) arrive at a hyperbolic functional relation $f = s\,a\,/\,(s + b)$ where f is flow, s is sunspot activity, and the parameters are a (the limiting value of flow) and b (the number of sunspots at half the limiting flow). A simple hyperbola rises and turns over towards a limiting (asymptotic) value. This is not unreasonable as the flow in a river is ultimately limited by the maximum possible rainfall on its catchment. There is uncertainty in both flow and sunspot activity, so the quantity to be optimised and the method for doing so are more complicated than in regression. The best fit to a hyperbola is added to Figure 9.4. The values we are interested in are the limiting flow $a = 26\,000$ m^3 s^{-1}, and the half maximum, $b = 22$ sunspots.

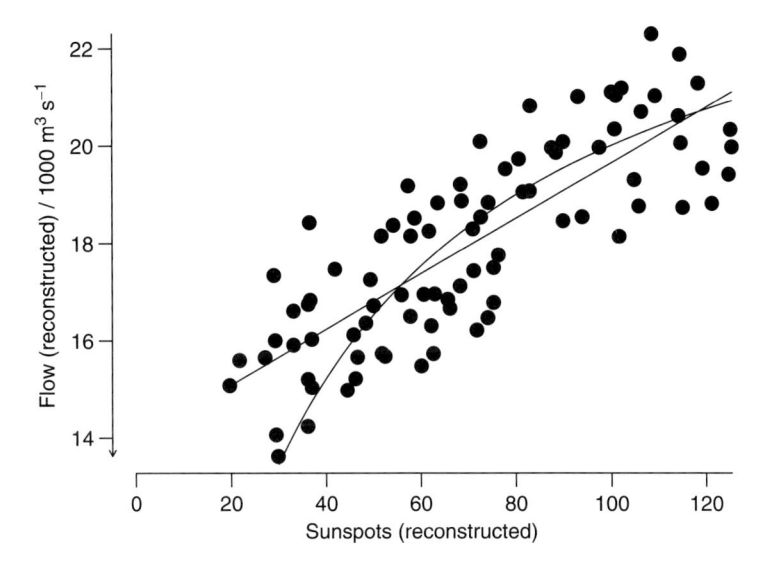

255 CORRELATION, REGRESSION, AND FUNCTIONAL ANALYSIS

FIGURE 9.4 Same data as Figure 9.3. The regression line $y = bx + a$ was used to predict two missing values of river flow from the known sunspot activity. The curved line is the fit to the functional relation $f = sa / (s + b)$ where f is flow, s is sunspot activity, a is the limiting value of flow, and b is the number of sunspots at half the limiting flow.

Of course this is only a skeleton of the analysis – we should need to know at least the confidence intervals (CI) for a and b.

Some workers see a scatter plot and (without much thought) reach for their linear regression program, mainly because they have seen others do so dozens of times. Sometimes they thus inadvertently use the correct calculation. But even then other pitfalls lie in wait. One of the commonest is illustrated in Figure 9.5, which shows the (simulated) mean mass of individual eggs, m, in nests of a species of bird in relation to the number, n, of eggs in each nest. It seems natural to look on the mean egg mass as dependent on the number of eggs: at least the reverse seems less natural. The number of eggs is known exactly and was 'chosen' by the (hypothetical) worker so the linear regression program can be used to make the functional calculations. There seems to be a negative linear slope. The more eggs the smaller, on average, is each egg. But stop and think for a moment. Suppose for each nest

256 DATA INTERRELATIONS

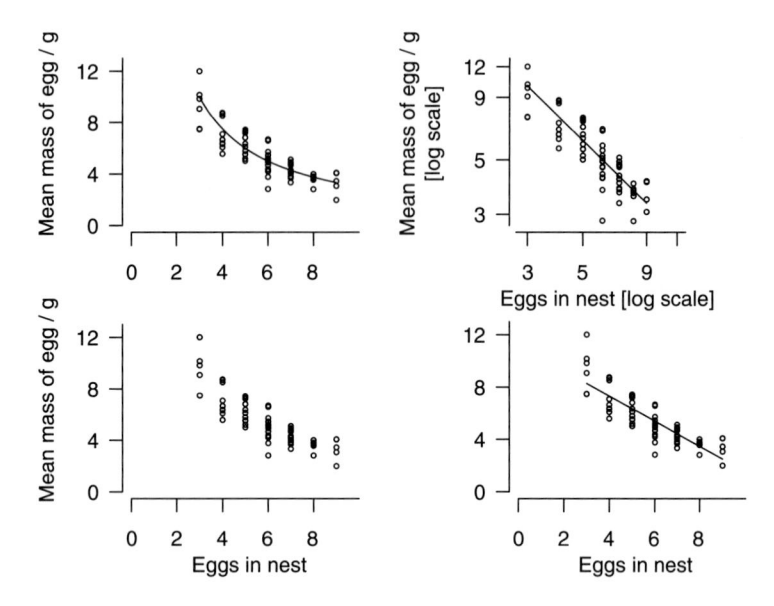

FIGURE 9.5 Mean mass of eggs in a nest dependence on number of eggs in nest. The data are simulated for a fixed total mass (30 g) with Gaussian error added (see text). *Lower left*: raw data. *Lower right*: with straight line fitted. *Upper left*: with reciprocal (the true relationship) fitted. *Upper right*: both axes log transformed with the theoretical straight line fitted.

the mother bird can provide a fixed total mass, M, of egg material but her physiology allows her to allocate that total to many small or fewer larger eggs. The functional relation would be $n\,m = M$, so we should get $m = M / n$, i.e. a graph in which mean egg mass is proportional to the *reciprocal* of egg number, not linearly proportional as the linear relation assumes. This arises whenever some underlying constraint (M) is split into two products (n, m) that are plotted on the two axes. The reciprocal is shown in the upper left graph in Figure 9.5. The true relationship is also shown by a straight line when both axes are log transformed (upper right graph). As Figure 9.5 shows, by the time some random scatter is added the true relationship is obscured. If you seem to have a linear relationship sloping downwards on a graph, it is worth asking yourself the question: 'Is it plausible that multiplying the variables on the axes should give a (roughly) constant value?'.

QUESTIONS:
- Have you published a correlation or regression?
- Was that the correct technique?
- Have you ever asked a question that requires functional analysis?

A better way?

The problems of null hypothesis significance testing (NHST) have long been recognised (for example, Läärä 2009) yet the procedure still dominates the bioscience literature, perhaps because it is easy to follow the procedure and there are numerous computer programs that will do the work without testing understanding. ▼ The arbitrary nature of $P \leq 0.05$, and risks of misinterpretation, have already been mentioned. Some of the other difficulties are these. An NHST encourages the TRUE or FALSE approach, which may be sufficient for a simple designed experiment. But it distracts attention from comparison with other models (not just a single alternative hypothesis). Very often we really want a quantitative assessment of difference while the null is just a single value at the far end of the scale. For example, suppose that the height of individuals in the population of two large towns is found to differ significantly. If the difference is barely 1 mm the interpretation is very different from that if the difference is 115 mm. With sufficient work one can creep below $P = 0.05$ for almost any null hypothesis but the result may be of trivial scientific importance.

> ▼ Several decades ago a statistician wrote a computer program to perform hypothesis tests, but only after the user had answered a short series of questions designed to assess the user's understanding and give helpful explanations when the answer was incorrect. The program was never popular.

As Sherlock Holmes remarked, "These are deep waters, Watson." There are methods that solve some of these problems, for example information-theoretic methods such as Akaike's Information Criterion, AIC (Akaike 1973, Motulski & Christopoulos 2004). The goodness of fit between data and a hypothesis can be measured by the likelihood of the data given the model (representing that hypothesis). But more complex models, with more parameters, can usually be tailored to give a better fit, and hence to produce a higher likelihood. The AIC measure of model quality compensates for this effect by requiring a better fit (higher likelihood) for a more complex model. In effect, extra parameters incur a penalty. The AIC thus allows a fair comparison of different models. We might have used this technique to indicate which of the two egg mass and numbers hypotheses is the more likely: linear (two parameters) or reciprocal (one parameter). The dimensionless AIC for the linear model is 6.6, and for the reciprocal model 4.6, indicating that the reciprocal hypothesis, with the smaller AIC, is the more likely to be true (which in this example, for which we generated the data from the 'reciprocal' hypothesis, we know *is* correct).

The most widely applicable alternative is to use Bayesian methods. Conventional (frequentist NHST and AIC) statistical methods have two major defects. Hypothesis testing asks 'Given the hypothesis, what is the probability of getting the observed data?' (D | H), though our real interest is usually the opposite 'Given our data what is the likelihood that a particular hypothesis / model is true?' (H | D). The other difficulty is that there is no way to include prior knowledge – information gathered from previous work or experience. This is often – perhaps usually – unnatural. You might have collected data about the number of boys and of girls in the population of a large town, expecting rather strongly on previous experience, that the quotient of boys / girls will be within 0.05 of 1.0. The pretence that you will accept with equanimity a value of 0.1 is just that: a pretence. Most investigations are not made in a vacuum. In 1763 there appeared, posthumously, a short article (Bayes 1763) by the Reverend Thomas Bayes, a nonconformist minister in Tunbridge Wells, that began a totally different approach to statistics.

Bayes discovered a way to include prior information in a formal way in an analysis. This information about the 'prior probability distribution' is used with 'new data' from your experiment or observations to give a 'posterior probability distribution' for the hypothesis / model.▼

> ▼ 'Prior' and 'posterior' refer to their place in the calculation: the prior may not come to hand until, during, or after data collection.

Many statisticians hold strong views for or against Bayesian statistics. It may be difficult to decide on the prior distribution; but if the distribution is near neutral – equal probability of all values – then the posterior distribution is largely determined by the data (and the results will usually be close to those given by frequentist statistics). There are many accounts of the Bayesian approach. An excellent explanation and discussion of the merits and drawbacks of NHST, information-theoretic, and Bayesian methods is in McCarthy (2007). I found Buck, Cavanagh, & Litton (1996) helpful too.

The characteristics of the three approaches are summarised in Table 9.3.

Why are Bayesian methods not more commonly used? They do require at least some mathematical understanding and programming skill, and need more computer time (decreasingly important), than do conventional methods. Before the 1980s the computations needed for a Bayesian analysis were not practicable for most bioscientists. The application of MCMC (Markov chain Monte Carlo) methods and the great increase in computer speeds transformed things. Computer applications that deal with common situations have appeared, and the use of Bayesian statistics is bound to increase as they become better known. One program written by (and intended for) statisticians is 'WinBUGS' (Windows Bayesian inference using Gibbs sampling). Others exist as packages for the 'R' general-purpose statistics and graph drawing program (see Chapter 10).

TABLE 9.3 Approaches to hypotheses and models

Criterion	Method		
	Frequentist		
	Null hypothesis significance testing (NHST)	Information theoretic	Bayesian
Number of hypotheses/ models	1	2 or more	1 or more
Prior probability Included?	No	No	Yes
Given	Null hypothesis	Hypotheses / models and number of parameters	Prior probability and new data
Test	Probability of getting these *or more extreme* data	Relative likelihood of the data for each hypothesis / model	Posterior probability of the hypothesis(es) / models

'Significance' and 'importance'

Remember that a result that is statistically significant may be of little or no scientific importance. Tests of statistical significance answer the question 'Is the effect real?' But beyond that is the question 'Does it matter?'

'Common sense'

Amid the complex, highly technical, specialist, hazard-ridden, and sometimes frightening processes of data analysis never forget or abandon the guiding light of common sense. 'It shall be a light to you when all other lights fail.' Are the results of an analysis credible? If not, is that because your expectations were wrong, or is there some mistake in the processing. Treat data-processing programs with respect, incorporating as they do the skills of experts, but remember that the programs are servants, not masters.

10

Tables and figures: the evidence

Tables list, and figures display, the evidence: text provides only the background narrative, and the interpretation. It is surely worth putting effort and time into displaying the evidence, on which the whole enterprise rests, as clearly as possible. Besides, after the Title and Summary, many (most?) readers scan the tables and figures next before deciding whether to read the full text – or not. Grammar, syntax, and the meaning of words govern the style of text; the styles of tables and figures derive from their design. For your first, or first few, articles you may be content with the standard tables and figures produced with a word-processor and spreadsheet program. But these programs are Procrustean – they force you to accept results that are less helpful to readers than they might be. If you are at all critically minded, and have pride in your work, you will want to improve the quality of your tables and figures, just as you begin to become interested in improving your text style.

Table 10.1 shows one way of grouping the main types of data: the more important distinctions are among continuous, discrete, and categoric types (which require different sorts of bar chart).

Compare the same data in Table 10.5 (later) and Figure 10.1. The content of these shows that authors in the general purpose journals *Nature* and *Science* use figures much more often than they do tables, as did the authors of the IPCC 4th report on the physical basis of climate change (IPCC 2007) who used 62 tables

TABLE 10.1 Types of data.

All data				
Quantitative = 'numeric'		Qualitative/categoric		
Continuous = floating point = real ('24.792')	Discrete = integer ('73')	Ordinal[a] ('Position = top, middle, bottom')	Nominal[b]	
			Binary ('Place = home, away')	Multigroup ('Party = Green, Liberal, Labour, Conservative')

[a] Ordinal data can be arranged in some natural order.

[b] Nominal data can usually be arranged in alphabetic order ('Conservative, Green, Labour, Liberal'), but except in dictionaries, telephone books, and so on, this is not considered 'natural'. Whether or not there is a natural order depends on the context. In the context of energy carried by a photon the colours red, green, and blue have a natural order and colour is ordinal, but a single pixel on a VDU has a colour that is a mixture of the three primary colours which for this purpose are nominal.

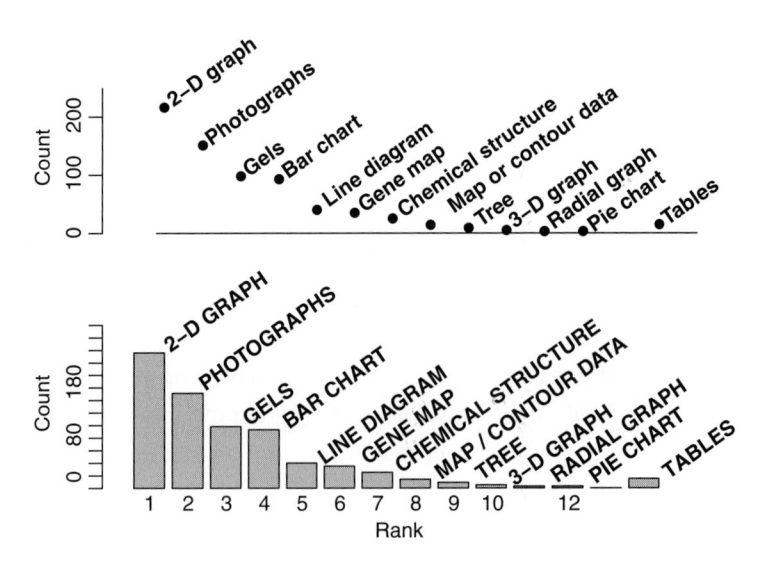

FIGURE 10.1 Number of graphic images of 12 types in one issue each of Nature and of Science. *Lower*: graph drawn with default settings. Note how the label for the first bar is truncated (it should read '2-DIMENSION GRAPH'). *Upper*: same data but with most 'chart junk' (Tufte 1986) removed.

and 252 figures (many of them with several graphs). Specific subjects, of course, have distinctive spectra of types of data presentation. Molecular biochemical data have eroded the distinction between tables and figures: for example one sees figures that include small tables or long nucleotide text sequences, and tables often contain small pictures of gels (and chemical structures).

Tables are generally better than figures at summarising information (particularly text, and especially for more than two or three variables) and at recording precise values that allow readers to make their own calculations.

Figures show patterns and spatial relationships among a few variables much better than tables do, and are usually the better choice in a talk or poster when the members of your audience need to grasp a pattern quickly. Figures are often better than tables at showing all the data (as in a scatter graph).

Tables

Tables are most often used for background and context information (for example site or experiment characteristics), for lists (such as ^{14}C dates in an Appendix) and summaries, and for some sorts of results (for example contingency data, and analyses of variance). In this chapter I have used a variety of table styles and formatting to illustrate some of the possibilities, but your target journal will probably use only one style. In particular, look for use of horizontal and vertical lines ('rules') and use of italic and bold face in column and row titles. Many journals allow no vertical rules, only sparing use of horizontal rules, and italic face only in titles.

Tables have five strengths:

1. They keep their contents together. Mendel listed, in a single text column, the number of peas of the four kinds resulting from what we now call an F1 cross. In translation his results were:
   315 round and yellow
   101 wrinkled and yellow
   108 round and green

32 wrinkled and green

This list might easily be broken by a page end, though I cannot tell as I write whether this will happen or not. With bad luck the two parts may even appear on backing (not facing) pages. This is a strong reason for making all such lists into independent tables that, provided they occupy less than a page, will not be broken. Mendel's results are shown in several tabular forms in Table 10.2.

2. Tables can show exact values, particularly of counts.
3. They avoid repeating text that sets each item in context.

TABLE 10.2 Number of each type of F2 peas in Mendel's experiment. The two characters recorded were round (R) or wrinkled (W) shape, and green (G) or yellow (Y) colour. The grandparents were pure-breeding of types RG and WY. All their progeny – the F1 – were RG. The F2 are the progeny of 15 F1 plants pollinated by other F1 plants. Three versions of this table are shown.

Top version: Default settings; columns too wide for easy reading (and wasteful of space) and no differentiation of headings. The word-processor default for tables is often unsatisfactory.

Lower left version: Columns only as wide as they need be to contain their contents on a single line; full 'furniture' vertical and horizontal dividing lines, and boundaries; numerical cell contents centred. This format is better but can easily be further improved.

Lower right version: Horizontal furniture only, and no boundaries; all cell contents centred. This is the format that many journals use.

Further emphases, if necessary, might be given with uppercase or italic font, or even by bold face.

		Shape	
		Round	Wrinkled
Colour	Green	108	32
	Yellow	315	101

		Shape	
		Round	Wrinkled
Colour	Green	108	32
	Yellow	315	101

		Shape	
		Round	Wrinkled
Colour	Green	108	32
	Yellow	315	101

4. Because they avoid repetitive text around each item, and arrange items in a regular pattern, they make comparison of items, particularly multivariate comparisons, easy.
5. They allow complex mixtures of text items and numeric items.

Most tables have four elements: heading (also called caption or legend), layout, 'furniture' (vertical and horizontal ruled lines), and data items. Many journals have specific styles for some or all these. The journal instructions may tell you what the styles are, but if not then scan recent articles in the journal to discover what is needed. A copy-editor may impose the journal format or style whatever you do, but will not often redesign the layout of a table to make it more easily comprehensible: that is your job.

Journals used to require that each table, however small, and its accompanying heading be on a separate sheet or page after the main text of what you submit, but this requirement is less common nowadays.

Table heading

I use 'heading' to stand for what can also be called 'caption' or 'legend', and I use 'title' for column headers and row labels. Each table must have an identity tag (for example 'Table 10a') and a brief description of what is in it that allows it to stand alone. Examples are 'Table 6. Dry bulk density of French cheeses.', 'Table 7.2b Primate Chromosome Numbers'. Inspect a recent copy of the journal to answer questions such as the following.

- Is the heading a phrase, with no terminal '.', or is it a complete sentence (which does have one)?
- Is the word 'Table' capitalised, or perhaps abbreviated to 'TAB.' or 'Tab.'?
- Is the initial letter of each word, or each major word, capitalised?
- Is the 'heading' to be above the table or below it?
- If the heading is above, is other explanatory information above the table (after the heading) or below the body of the table?

There is general agreement that a table heading should not interpret or comment on the table contents: avoid 'Negligible percentage of Martians landed in the night'. Some editors think that in table headings participles are preferable to relative clauses (introduced by 'that' or 'which') – 'Martians landing in the night' is better than 'Percentage of Martians that landed in the night' – but there is no agreement about this. The second version has the advantage of specifying the nature of the cell entries ('percentage') and if all the cell entries are of the same kind this displays the units in a prominent place.

Most authors and editors do seem to agree that a table heading should start with a short phrase or sentence describing the contents. But should one follow that, where necessary, with background information? Advocates of the single short one-phrase (or one-sentence) heading believe that other relevant information will be shown in column or row titles, or in footnotes. But tables in scientific reports often do require extensive background to be interpretable. That the contents of the table were 'gathered at the Cumberland site, in summer, by group B of three observers, between 9:00 and 11:00, on a rainy day, by method R20' does not fit into a snappy one-phrase heading but is important if the table contents are to be judged accurately. The obvious place for such information is just after the initial heading.

Do not put into the heading details that can or must be repeated in column or row titles: 'Effect of nitrogen additions at 10, 20 and 40 g m^{-2} on growth after 3, 9, and 20 weeks' can be reduced to 'Effect of nitrogen addition on growth' with the details incorporated in the column and row titles.

Column and row titles are often similar to graph axis titles: 'Rate of uptake of CO_2 / mmol hr^{-1}'. They are frequently badly composed, as described in Chapter 7: 'Expressing the name and value of a variable'.

Table layout

Include only necessary data in tables: omit everything that is not relevant to this particular article.

In Table 10.2 (above) each version takes the same number of lines as the original text list. Round and wrinkled shape is separated horizontally (in columns), while colour is separated vertically (in rows). In this example shape and colour could be interchanged: the arrangement shown in Table 10.2 is arbitrary. In the same way the choice of graph axes when a correlation is of interest is usually arbitrary.

In tables that show background information or results about, say, sites or experiments (I will call them 'instances'), the preferred arrangement is the instances down the page as rows and attributes as columns across the page so that cells the reader will most probably want to compare are in the same column. Why? Because it is easier to compare items in a column than in a row. Another test is that items that may sensibly be added together should be in the same column. But there are other, and possibly more important, considerations too.

If a table is to allow the user to locate a named item then the rows are best ordered alphabetically (as in a telephone directory) but usually the interest will be in the numerical values in one or more columns. Suppose, for illustration, that the first column contains the date of observation, the fourth contains temperature, and another shows the number of pupae hatching. Initially the rows are in date order, because that is how you entered them in the computer. But if your main interest is in the effects of temperature on hatching then it would help readers if you rearranged the rows in order of temperature.

A desirable objective is that the table should be oriented so it can be read without turning the journal round physically into the landscape position. (Journals are printed to be read in the portrait position, ▯, with height greater than width. In the landscape position, ▭, the width is greater than the height.) Most journals now have more than one column of text on a (portrait-oriented) page, and tables can usually be printed to span one, two, or more text columns up to full page width. (Some journals will even print a table across two facing pages if necessary.) You should design your table to span a specific number of the text columns. The easiest way is probably to create the table, then to adjust column widths in it, and perhaps font sizes, to achieve the best arrangement, then relax a bit to fit the table to the number of text columns on the printed page. Particularly difficult are table columns with long titles but short entries. In Table 10.3, showing a complex set of data, the column titles have been spread over several lines, and ancillary information relegated to notes at the bottom of the table.

Table 10.3 has to accommodate 10 different attributes as columns and 14 instances (solar system bodies, SSBs) as rows. The table also uses column and row spaces to separate titles. But the SSBs have many more attributes than are shown in Table 10.3, and if a few more were to be included the columns would not fit into a single page width. There are two solutions. (1) Keep the same orientation and make several tables, repeating the SSB names in the first column of each table. (2) If there are not too many SSBs to fit across the page, then put them as the columns and the characteristics as rows. This solution requires a reader to make comparisons along a row, and is thus the less desirable choice.

The improvement in understandability that can be got by redesign is shown in Table 10.4. The first version (10.4a) is hypothetically as supplied by the authors of a submitted article. It could be accommodated on the page only in the landscape orientation, and to fit it to the portrait orientation I have abbreviated the column titles. The interest is in three comparisons: geographic, aspect, and topographic (hummock or hollow). That is not easy: the variables are not simply instances or attributes. The important values are the means, while the standard errors are subsidiary. In Table 10.4b they have been placed inconspicuously as subscripts. Italic face is used to help make comparisons. The whole table now allows three sorts of comparison: concentrations decline from west to east, hummocks have larger concentrations than hollows, and (with one small exception) concentrations are larger on sloping than on flat mires. Because there were at least

TABLE 10.3 Some characteristics of the Solar System. This is a complicated table with text and numerical entries, columns of different width, footnotes expanding the column titles, and other footnotes of different kinds in the row titles and in some of the cells. Italic and bold face are used judiciously to draw attention to a few entries. Some column titles are much longer than the entries so are formatted as more than one line.

Body	Equatorial radius[1]	Density / Mg m^{-3}	Mass[2]	Surface gravity[3]	Orbit[5] around sun / AU[4]	n	B (Bode's Law)[6]	Number of moons	Largest satellite
Sun	109	1.41	333 000	28.0					
Mercury	0.38	5.43	0.055	0.38	0.37	1	0.4	0	-
Venus	0.95	5.24	0.82	0.90	0.72	2	0.7	0	-
Earth	*1.00*	*5.52*	*1.00*	*1.00*	*1.00*	3	1.0	1	Moon
Mars	0.53	3.94	0.11	0.38	1.52	4	1.6	2	Phobos
Ceres*†	0.04	2.08	0.0002	0.03	2.77	5	2.8	0	-
Jupiter**	11.2	1.33	318	2.54	5.20	6	5.2	63	Europa
Saturn**	9.4	0.70	95	1.07	9.54	7	10.0	60	Titan
Uranus**,‡‡	4.0	1.30	14.5	0.9	19.2	8	19.6	27	Titania
Neptune**	3.8	1.76	17.2	1.2	30.1	9	38.8	13	Triton
Pluto*‡	0.18	2.01	0.0025	0.07	39¶			3	Charon
Haumea*‡	0.18?	3?	0.0007?	0.05?	43¶			2	Hi'iaka
Makemake*‡	0.12?	2?	0.0007?	0.05?	46¶			0	-
Eris*‡	0.21?	?	0.0031?	0.08?	68¶			1	Dysnomia

Note: many of the values for the three outer minor planets (Haumea to Eris) are uncertain.
[1] Equatorial, *relative to Earth* (12,756 km).
[2] *Relative to Earth* (5.976 x 10^{24} kg).
[3] *Relative to Earth* (9.78 m s^2).
[4] *Relative to Earth* (149.6 Gm) = 1.0 Astronomical Unit (AU).
[5] Orbit semi-major axis.
[6] Bode's Law: $B_1 = 0.4$; $n = 2$ to 10, $B_n = 0.3 \times 2^{n-2} + 0.4$. Note the correlation of B with the semi-major axis out to Uranus.
* These five are considered to be minor planets: they are in orbit around the sun, in hydrostatic equilibrium (mostly approximately spherical), but have not cleared their orbits of other bodies.
† The largest object in the asteroid belt. This belt contains tens of thousands, possibly millions, of mainly rocky objects more than 1 km diameter, similar to the four inner planets, as well as uncounted smaller bodies. It is thought that gravitational interference from Jupiter prevented these bodies from coalescing.
** The four 'gas giant' outer ('Jovian') planets that together contain more than 99% of the known mass of the solar system outside the sun.
‡‡ Uranus orbits the sun on its side.
‡ Trans-Neptunian belt. Contains uncounted objects, mainly of rock and ice.
¶ Orbit is strongly eccentric.

40 measurements in each sample there is not a lot to be gained by calculating confidence intervals (CI), but had there been fewer than 25 measurements in each sample then CI would be better than SEM (Chapter 8).

TABLE 10.4 (a) Na$^+$ concentration / mg l^{-1} in water from hummocks (Humck) and hollows (Hollw) on flat and sloping peatland in three geographic regions. Values in parentheses (in the lower row) are the standard error for $>= 40$ measurements.

The original table was submitted in landscape orientation with each column about 60% wider than shown here. It has been necessary to abbreviate 'Hummock' and 'Hollow' to get the table to fit this page in the portrait position.

(b) Na$^+$ concentration / mg l^{-1} (subscript is standard error, $n >= 40$) in water from hummocks and hollows on flat and sloping peatland in three geographic regions. *Hollow* values are italic for clarity.

Rearranged from Table 10.4a. The original layout needed row and column lines to be intelligible; the rearranged one does not, and the standard errors are less prominent. 'REGION' is uppercase to make clear that it applies to its three subdivisions in the line below. It is easier now to see that concentration decreases from west to east, that it is less in hollows than in hummocks, and that it is generally less on flat peatlands than sloping ones.

(a)

Western flat		Western sloping		Central flat		Central sloping		Eastern flat		Eastern sloping	
Humck	Hollw	Humck	Hollw	Humck	Hollw	Humck	Hollw	Humck	Hollw	Humck	Hollw
8.9	5.8	13.8	9.9	4.0	2.3	6.5	2.2	1.6	1.4	2.1	1.7
(0.9)	(0.3)	(0.5)	(0.2)	(0.4)	(0.3)	(0.6)	(0.3)	(0.1)	(0.1)	(0.2)	(0.1)

(b)

	REGION					
HABITAT ▼	Western		Central		Eastern	
	Flat	Sloping	Flat	Sloping	Flat	Sloping
Hummock	8.9$_{0.9}$		4.0$_{0.4}$		1.6$_{0.1}$	
Hollow	*5.8$_{0.3}$*		*2.3$_{0.3}$*		*1.4$_{0.1}$*	
Hummock		13.8$_{0.5}$		6.5$_{0.6}$		2.1$_{0.2}$
Hollow		*9.9$_{0.2}$*		*2.2$_{0.3}$*		*1.7$_{0.1}$*

Table furniture

Some journals will accept tables as spreadsheets, and will then arrange the furniture themselves. Usually, however, you have to construct the table yourself. Inspect a recent copy of the journal to answer questions such as these:

- Is each cell in a table bounded by ruled lines?
- Do tables have vertical ruled lines to separate columns and perhaps to indicate column grouping?

- Do they have horizontal ruled lines to separate rows and perhaps to indicate row grouping?
- Do they have a boundary box?

A general rule is to minimise furniture, especially vertical lines, but complex or large tables may need a few lines to help the reader group columns (and rows and cells) correctly. In this book I have frequently tried to keep table columns and rows to the minimum width, and this has required either vertical or horizontal rules, or both, to separate items clearly. A journal would use greater widths to minimise the need for furniture. Merged cells in the body of the table (Table 10.1, for example) may also necessitate more furniture. Groups of rows can be separated by an extra line space, and groups of columns can have a group title centred over separate column titles below it, or a group title that is extended before and after the text with dashes to span the whole group (Table 10.2). In that table the first version, produced with my word-processor's default settings, has unnecessarily wide columns which waste a lot of space, and make it difficult to align items along the rows. The attempt to show the group spread of 'Shape' is crude. Nor are the group and column or row titles differentiated. The other versions correct these defects and show the use of vertical, and horizontal lines, and a bounding box, with differently differentiated group titles and data. The last version has no bounding box, and the complex Table 10.3 contains no rules (lines) at all. It is worthwhile learning how to control the word-processor's table settings. By 'selecting' parts of the table you can apply, for example, lines or centred values, to the selected part alone.

Data items in a table

Think about the units you use. A long column of mass values reported in g and all beginning 0.00 ... is difficult for your reader. Change the units to mg and all the 0.00 are unnecessary. Whatever you do, avoid repeating the units with each value in a table: give the units once only, in the row or column title (or the table heading if all the cell entries have the same units).

Get the precision of numerical values right: report only *significant* digits (Chapter 8 explains how to determine what is significant). Where precision differs within a column use different numbers of digits after the decimal point. For the reasons given in Chapter 8 resist any editorial attempt to give every value in a column the same number of digits after the decimal point: it is your reputation that will suffer.

There are four types of special entry. (1) The item is logically impossible, for example the name in Table 10.3 of the largest satellite of a planet that has no satellites. A dash ('–') or simply a blank is usually used. Table 10.3 uses both for slightly different purposes (to distinguish the Sun from other bodies). (2) An item could in principle exist but for one of uncountable reasons is not known: for example the value was lost, it was never measured (accidentally or unavoidably), the measurement was later found to be unreliable. The entry 'n.d.' or 'nd' standing for both 'no datum' and 'not determined' is usual. The entry 'n.a.' or 'na' for 'not available' is sometimes used, but suggests a small mystery. Was it measured but then suppressed? If so, why? (3) The value was measured but was less than the limit of detection. For this the entry 'nd' for 'not detectable' is sometimes used but should not be because it is logically different from 'not determined'. Much better is '< 0.01' or whatever the limit of detection was. In any case this entry is more informative as it specifies what the detection limit was. (4) The counted or measured value is zero. If there are few zeros then they are best shown simply as '0' (integer) or '0.0' (continuous). But if the values are counts and many are zero then they can be shown less obtrusively by '...', as they are in Table 8.1c.

Finally, compare the same data shown in Table 10.5 and in Figure 10.1 (a luxury you would not be allowed in a journal). The data are in rank order and it is clear in the table that there are big differences between those items with three digits and the others. But the figure shows the pattern much more clearly: the data fall close to a negative exponential (so when the ordinate is

TABLE 10.5. Frequency of kinds of evidence (top sub-table) and of 12 types of display in figures (lower sub-table) in one issue each of *Nature* and of *Science*. Most figures include several types of display. The information in the lower sub-table is also shown in Figure 10.1.

	Tables	Figures	Equations
Instances	15	135	20
% of total instances	6	82	12

Illustration type	Number	Illustration type	Number
(1) Two-dimension graph	216	(7) Chemical structure	25
(2) Photograph*	151	(8) Map†	14
(3) Gel	98	(9) Tree diagram	9
(4) Bar chart	93	(10) Three-dimension graph	5
(5) Line diagram	40	(11) Radial graph	3
(6) Gene map	35	(12) Pie chart	3

* Excluding gels.
† Geographic, and data contour.

log-scaled they are close to a straight line). Figure 1.1 (journal impact factors) showed a similar pattern. These figures then prompt a question about the functional cause of this pattern when a set of data is divided into categories. That question does not emerge from a table.

It is true that the exact values are more difficult to get from figures, but with email near universal among scientists there is little difficulty in asking for raw data from the author (the response is unpredictable).

Figures: graphs

The commonest element in a figure is a two-dimensional graph. Graphs are good at showing relationships ('messages') amongst numerical values that may be numerous and in complicated groups. Always examine results graphically if possible. This first view, before the thought and analysis that follow, need not consider details of the design of graphs. Indeed many authors get no further at all and publish, often ill-adapted, standard graphs that a spreadsheet program has produced. If the article incorporates six months work then surely it is worth spending half a day presenting the graphs well for the final publication?

The purpose is, as before,

to make your readers' task as easy as possible.

As with text and tables you need to consider style and design details of several kinds. What follows is mostly about producing graphs for publication. The journal editor has firm control of text style, some control of table style, but much less of graphics style, so whatever you choose will probably pass through the editorial process almost unchanged, as a cherry stone does through a bird.

The technical skills needed for text, tables, and figures form a series from easy to more difficult. Most writers produce text with one or other of a small number of word-processors and these are easy to learn to use. The same programs will produce tables, but learning to control the details is more difficult. The same 'office suites' do have primitive graph-producing abilities but these are near useless for those with any pride in their work. However, there are numerous free-standing programs that can produce a wide variety of graphs, with differing degrees of control of details. Learning to use such a program needs a noticeable but unavoidable investment of time, so (if you have not already done so) it is worth surveying the possibilities ▼ before making a choice.

▼ Two programs that I have found useful are Sigmaplot (expensive, controlled by mouse and keyboard, runs on MS Windows only); and 'R' (free, controlled by typing commands, runs on Linux, Apple, and MS Windows). There are many others. Many journals specify the file format in which figures should be submitted. Sigmaplot and 'R' are large and powerful data-analysis packages with closely controllable graph plotting. I found 'R' the hardest, by a little way, to learn. Sigmaplot includes a manual. 'R' has extensive on-line primers and help, and several books sold separately. Helpful for a beginner are Knell (2013) and Venables *et al.* (2009).

All the figures in this book, except 3.1, were made with 'R'.

When you see a graph for the first time you use two different mental processes. First is pattern detection, viewing the graph as a whole (for example, are those two clouds of points close to two curved lines that cross?, are the red points generally below the blue ones?) followed by analysis of detail focusing on specifics (for example, what are the axes?, are the axes linear?, what do the different data symbols mean?).

Design: general

Tufte, in two influential books (Tufte 1986, 1990) advocates using only essential marks – he calls everything else 'chart junk' – and treating ink as if it were as valuable as liquid gold: maximise information, minimise ink. In the days when one drew graphs with ruler, flexible curve, template, pen, and ink, and perhaps stick-on letters, then it was natural to omit redundant marks, and one could just as easily draw an unusual design as a standard one. But the writers of many graph-drawing computer applications seem to be ignorant of needs and conventions; their defaults seem to be chosen to show off what is possible rather than the minimum that is necessary; and many programs are too inflexible to allow you to draw exactly what you want. This is another reason for choosing carefully the application program in which to invest time.

Put in the graph only relevant data, and strip out everything that does not help the message. For example, Tufte suggests that you do not box the graph (unless the journal requires it); minimise the number of ticks and labels on axes; avoid unnecessary colour (grey is a colour). Figure 10.1, upper, shows a Tufted version of the same data as Figure 10.1, lower. The ticks and labels on the abscissa were superfluous. The ordinate now has only three ticks and labels (enough to confirm that the scale is linear). The annotations use an initial capital only. The graph is unboxed. This approach may be taken too far (though I know of no examples except those designed to show the problem): unfamiliar sparse forms can impede understanding. In particular some scientists do like boxed graphs (though these have other disadvantages, see later). Balancing data density and complexity, clarity, simplicity, and aesthetic appeal is often difficult, and is a cause of conflicting advice.

Before starting detailed design consider these things.

- Decide what size to design at. First, look at your target journal and decide what final size to aim for. Is your graph a simple one, so that a single print column will be sufficient, or is it more complicated needing two or more columns? Editors try to reduce the number of separate figures by requiring, where sensible, more than one graph in a figure, with a single modestly extended caption (Figures 10.1, 10.2). Can you anticipate this demand? Can you use a simple smaller inset graph subsidiary to a main one (Figures 1.1, 9.1) rather than having two separate figures?
- Choose the aspect ratio – the length : breadth ratio – so that as far as possible the slopes of neighbouring points or their trend lines are near 45° (Figure 10.2, upper, contrasted with the same data in Figure 10.2, lower right) because near this value human judgement of slope is most acute. The editor may be grateful to know what size your design is intended to be when printed.
- The journal may be able to shrink or enlarge what you submit or may require you to supply at the final size. If you design at twice or more the intended published linear size, then small defects will be easier to see. A good graph-plotting program will allow you to change the image dimensions easily. Some graph-plotting programs have a fixed font size; others allow you to change the font size independently from the rest of the graph; some will allow you to keep the same proportion between font size and the whole graph whatever size you submit at. Whatever the requirements, when the graph is complete print it at the intended published size

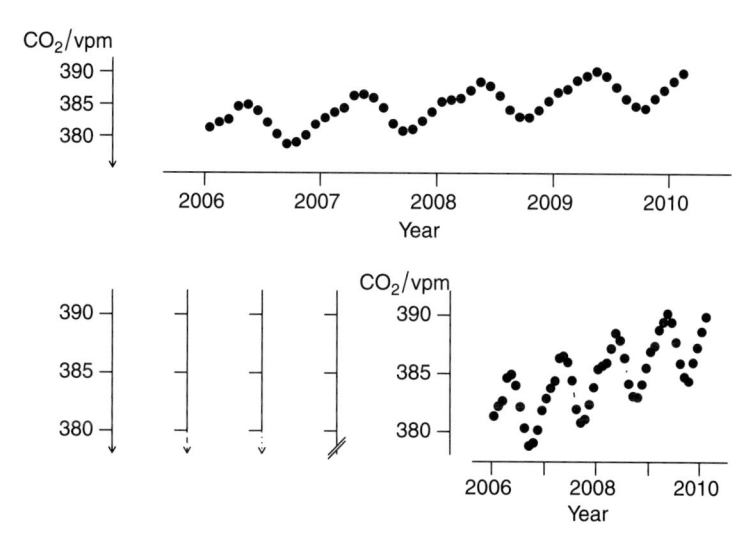

FIGURE 10.2 *Lower left*: four ways to show that an axis is truncated. *Lower right*: recent values of mean monthly CO_2 concentration in the air on Mauna Loa in Hawaii. The ragged ordinate does show that values below those plotted are possible but it does not draw attention to how distant is the theoretical lower limit of zero: the range of the plotted values is only 1/35 of their mean. Viewers may be alerted to this by one of the four conventions to the left. Note also that the tick labels are horizontal, and the axis title is above the ordinate. *Upper*: the same data plotted so that the slopes are about 45° to the horizontal, and with attention drawn to the suppressed zero on the ordinate. This plot shows that the pattern is asymmetric with a longer up-slope than down – a feature not obvious in the lower graph.

to check that details, line thickness, and text size are all sufficiently prominent. Check that you can also create a file in a format the journal will accept.

■ Most journals used to require figures to be black on white, with a large charge, perhaps 100 €/£/$, for colour. This may still be true for print on paper (it depends on the journal), but if the article will be put on a website by the publisher then colour will cost nothing. You may be allowed to provide a plain version for printing and a coloured one for the website. But whatever the circumstances it is salutary to try to design for black on white (perhaps allow grey tones if essential), and treat colour as a bonus if it is available. If you do use colour try to avoid different colours that have similar density because, when copied in black-on-white, differently coloured features may not be distinguishable from one another.

Arrangement of multiple graphs

If you need to put numerous graphs in a single figure, and particularly if they form a natural matrix, then it may seem natural to scan from the top left graph across and then down, as text is read in Western writing. But the individual graphs are naturally scanned from the axis origin, usually at the bottom left towards the top right. If the figure is to contain more than one but not 'numerous' graphs it may seem natural to ask the reader to scan the figure as one would a single graph from the bottom left to the top right. This is particularly the case if one graph is 'main' and others subsidiary. It is the arrangement I have used in many figures in this book, so you can judge its effectiveness.

Design elements

Graphs have style elements at several levels, just as text and tables do. Particularly helpful, though not in agreement, about design are Tufte (1986, 1990), Cleveland (1994), and Frankel (2002).

Be aware that elementary visual tasks differ in accuracy (Cleveland & McGill 1985). Comparisons, in descending order of accuracy are: position along a common scale > position on common but non-aligned scales > lengths > angles or slopes in the range 30 to 60° > areas > volumes, density and saturation of colours. In what follows I consider first: axes, data symbols, lines and annotations; and then some common types of graph.

Axes: general
AXIS ORIENTATION
Convention requires that where one variable is in some sense physically or notionally (as in regression) dependent on the other then the 'cause' be on the abscissa (horizontal axis) and the dependent variable on the ordinate (vertical axis). Exceptions are

when the independent variable ('cause') is naturally measured in the up-down direction, as for height, depth, altitude, which can be put on the vertical axis (Figure 10.5, upper right, later). Other special instances are stress–strain diagrams (where by convention the causal stress is often put on the vertical axis and the resultant strain on the horizontal axis), and needs such as keeping long labels horizontal on an independent axis. Where no causal relation is implied (correlation) then the allocation of variables to axes is arbitrary.

AXIS ENDS

Many variables have physical or logical limits: mass cannot be negative. For these, make sure that there is a tick mark at 0.0 and that the axis is not left projecting raggedly into the impossible or invalid region. But if a variable (such as mass) could, in principle, have been larger than any measured value then it is helpful to let the axis project raggedly beyond the last tick mark. Figure 10.2 shows examples. It is particularly helpful to observe this rule where a percentage may, in principle, be above 100 as for example where results are expressed relative to a control.

Some scientists like to include axes with ticks on all four sides of a graph ('boxed') arguing that this makes coordinates measured from the graph more precise. In practice this seems to me to be illusory. It is easy to photocopy and enlarge a published graph, and to measure with a set-square and ruler on this, or to use a digitiser. And you can often ask the author for the specific values. If the boxing has been done so that all axis intersections are completed then one cannot apply the advice about ragged axis limits.

ZERO SUPPRESSION

In the commercial world an increase in profits from 2.20 to 2.23 M£ will often be plotted on a graph extending from 2.15 to 2.25 M£ probably with the intention of grossly exaggerating the improvement. One hopes that no scientist would mislead deliberately in this way, but many do so inadvertently. The general problem is that of the 'suppressed zero'. Of course, anyone who reads the tick labels should realise what has been done, and leaving the axis ragged at the lower end will help. But if the distance to zero from the lowest ordinate is more than, say, the range of the ordinate data, then it may be worth trying to draw attention to the suppressed zero. This can be done in several ways: you can describe it in the caption ('Note the suppressed zero'); or you can use one of the modifications to the lower end of the axis shown in Figure 10.2, lower left.

TICKS

You need to include sufficient ticks and their numerical labels to make plain the scaling (linear or non-linear), but in most cases three or four ticks are sufficient: more are distracting. The commonest non-linear axis scaling is logarithmic, but many others are possible, for example 'hyperbolic', 'reciprocal'. These can be identified in the axis title (see later). Ticks on logarithmically scaled axes should be arranged roughly equidistantly, and may need to be more numerous than on a linear scale. For example over two decades of a logarithmically scaled axis you may tick and label 1, 2, 5, 10, 20, 50, 100. (See later for why you should use values in the original units and not their $\log_{10}$ (0.0, 0.3, 0.7, 1.0, 1.3, 1.7, 2.0). A compromise (for example, Figure 8.4) is to label ticks at 1, 2, 5, 10, 100 and to add unlabelled ('minor') ticks at 20, 50, 200, 500.

Where to position the ticks across the axis is usually an aesthetic decision. The main possibilities are projecting outside the data area, centred across the axis, and projecting into the data area. Outside leaves the data area uncluttered and avoids possible coincidence of a tick with a datum symbol; inside leaves less clutter among the axis labels. If the journal has no convention, I prefer to put ticks outside the data area.

AXIS TITLES (LABELS)

Titles appear on the axes of graphs (and as titles of columns and rows in tables). These titles have four potential elements. (1) The name of the variable. A phrase, with the conventional initial capital but lowercase thereafter (except where ordinary text would require a capital): for example 'Rate of uptake of ^{14}C', 'Spatial density of crocodiles' (not 'Density Of Crocodiles' or 'DENSITY OF CROCODILES'). Some journals insist on an initial lowercase letter: 'spatial density of crocodiles'. (2) Symbol, if used to represent the variable in the text: 'Rate of decay, ε / yr^{-1}'. (3) Units, if any, for the variable. (4) On graph axes: the axis transform, if not linear ('log' is the commonest).

The name or symbol is essential, and so are units if the variable has physical dimensions, but the symbol and scale transform are often unnecessary. Journals differ in how they expect or require axis titles (and table row and column titles) to be punctuated. In Chapter 7: 'Expressing the name and value of a variable' I argued that the best is the 'slash' convention:

'Name of variable, symbol / units [transform]'

but many, perhaps most, bioscience journals use an inferior parenthesis convention:

'Name of variable, symbol (units) [transform]'

The logic of the first form is that the value of a physical variable (which has two components: numerical value and units as essential parts of it) when divided by its units gives a dimensionless number. It is this pure number that is shown in the tick labels and is used to position the data points.

AXIS TRANSFORMS

I use the logarithmic transform of an axis for illustration, but the same principles hold whatever the transform. What may cause you to use such a transformation? First, the range of data may be large and the data become more sparse towards the upper end of the axis so that on a linear axis most of the points cluster and overlap at the lower end with rather few scattered towards the other end. Figure 1.1 is of this kind, the log transform puts the data points at roughly equal distance from each other on the plot. Second, you may have some underlying hypothesis that predicts that a logarithmic transform will convert a curve to a straight line. Growth and decay commonly do this. It looks as if Figure 10.1 might respond in this way (though it is not overtly based directly on growth or decay). Third, ratios (strictly 'quotients') are considered later and should usually attract a log transform of the axis.

The crucial point is that the labels on ticks should be in units of the original variable before logs were taken. There are two reasons: first the practical one. It is much easier for a reader to understand the original units than their logarithms. $\log_{10} = 1.2$, for example, is just about understandable as 'a bit more than 10 and less than 20' (which would be 1.3010). But what do you make of $\log_e = 1.2$ (or $\ln = 1.2$)? Some bioscientists carry a few values of the $\log_{10}$ scale in their memory, but how many do the same for $\log_e$? And what do you make of '$\log = 1.2$', with no base named? If the ticks and their values are at approximately equal log distances (as advocated above) then the values will make plain that a transform is being used.

That ought to be convincing by itself but, if you use the less satisfactory parenthesis convention for axis titles (for example 'Mass remaining (g dm^{-2}) [log scale]'), there is a second important conceptual difficulty. You could take the log of 1.56 (0.193 in base 10, 0.445 in base e, 0.642 in base 2), but how do you take the log of 1.56 g dm^{-2}? There is no way to find the log of a physical unit. What we are trying to do is to transform the *axis*, not the values, by stretching it like a sheet of rubber in a carefully graduated way more in one place than another. That leaves the original numerical values stretched to differing degrees but otherwise unchanged. If you plot such a graph by hand you calculate the log of each numerical value to show you where to put it on the differentially stretched scale. But once that is done you replace the log values on the axis with the corresponding original ones. The process resembles putting up a scaffold to build a house: when the house is complete you remove the scaffold.

If you use the better, 'slash', convention for labels then it is valid to have, for example, '$\log_{10}$ (mass remaining / g dm^{-2})' as long as you include the base of logarithms. But it still leaves readers to make the conversion from logs to original units, and for that reason is unhelpful.

It will probably be clear that the *shape* of the plot, the relative position of points on it, will be the same for all bases of logarithms, so the '[log scale]' does not need to specify the base.

Even worse than labelling ticks with their log value is an axis labelled as 'Log ... (g dm^{-2})'. The missing base does not affect the shape and distribution of points, but without specifying the base how is the reader to convert 0.45, for example, on the scale to the original units? Does it represent 2.8, 1.57, or 1.37 g dm^{-2} for the bases 10, e and 2 respectively?

If both axes are log-scaled then consider whether or not a tenfold range on one axis should occupy the same distance on the graph as a tenfold range on the other axis. Figure 8.4, upper part, has both axes transformed. There is a good reason for making the cycle length (tenfold change in numerical value) the same on both axes so the slope of −0.5 is obvious.

AXES FOR QUOTIENTS (RATIOS)

Graphs of quotients ('ratios') require thought. Figure 10.3 shows the same four hypothetical quotients plotted on linear and on log-scaled axes. On the linear scale at the left one's first impression might be that there is little difference between 'a' and 'b', and that the difference between 'A' and 'B' is very much bigger than that between 'a' and 'b'. That would be misleading. This is a result of cramming all possible values between 0.0 and 1.0 into the restricted linear space between 0.0 and 1.0, while having infinite space for values > 1.0. The value of 'b' is twice that of 'a', while the same is true of 'A' and 'B'. Indeed 'A' and 'B' are what one would get by simply reversing the order of the two elements in 'a' and 'b'. For example if, instead of the conventional C : N, one were to plot N : C. These things becomes obvious in the right side of Figure 10.3 where the values are plotted on a log-transformed scale with 1.0 as the reference value. The length of the 'a' and 'A' bars is the same, and so is that of 'b' and 'B'. A log-transform axis should usually be used if the second element in the quotient is being used as a

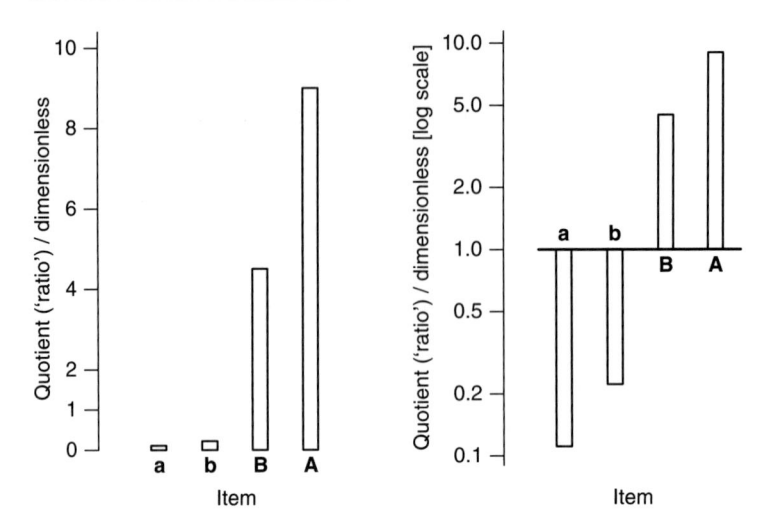

FIGURE 10.3 Four hypothetical quotients ('ratios') of value 'a' = 1/9 = 0.1111, 'b' = 1/4.5 = 0.2222, 'A' = 4.5, and 'B' = 9. Thus 'a' and 'b' are the reciprocals of 'A' and 'B' respectively. Plotted at the left on a linear scale, and at the right on a log-transformed scale.

reference to standardise the quotients, and always if some of the quotient values are less than 1.0 and some are more than 1.0. If the quotient values are all much greater than 1.0, as they would usually be for C/N values, for example, then the log transform is less necessary, but even in these circumstances it will do no harm.

The fundamental point is that quotients involve a division so, just as the now obsolete slide rule converted multiplication and division into simple linear addition and subtraction of lengths, so the log transform of the axis displays the quotients in a simple additive/subtractive way which humans find easy to interpret.

AXES: MISCELLANEOUS

Care is needed where a graph is to show a rate measured at a consecutive series of times. As example take rainfall measured in the successive years shown on the abscissa. The ordinate label may be 'Annual rainfall/mm'. But what is actually being plotted is a rate and the time belongs in the units. Yet it is also relevant that the values of the rate of rainfall have been averaged over particular years shown on the abscissa. The label should then be 'Annual rate of rainfall / mm yr^{-1}', the 'annual' indicating that this is the averaging period corresponding to the bar width on the abscissa.

In general, make labels as specific as practicable: 'Time after addition' not just 'time'.

Journals differ in the way that they format units and symbols. For example, some would specify 'mg / g / hr'. This is incompatible with the 'slash' convention for axis (and table column or row) titles, and is potentially ambiguous: it may mean either '(mg / g) / hr' or 'mg / (g / hr)'. That these are different is obvious when one uses specific values for example 32 mg, 4 g, and 2 hr. The first interpretation gives (32 / 4) / 2 = 4.0 mg g^{-1} hr^{-1}, and the second gives 32 / (4 / 2) = 16 mg hr g^{-1} (different units). 'Common sense' suggests that the first interpretation is the one intended, and the left-to-right rule for arithmetic evaluation agrees. But this rule is not universal, and one should not have to assume or guess what an author means. The problem is avoided if parentheses are used, but these might be confused temporarily with the parentheses used to surround the units in the less satisfactory but common parenthesis convention for titles. A simpler way of avoiding these problems is to use index notation: 'mg g^{-1} hr^{-1}'. This is particularly useful if the 'slash' convention 'name of variable / units' notation is used because the '/' is used for that purpose alone.

Consider the orientation of tick labels and titles on vertical axes. Parallel to the axis, which is often the default, is tolerable in print though it requires twisting one's neck or turning the page round to read the label and title. There is usually no reason why the tick labels should not be rotated to be upright. For a talk or poster a title for the vertical axis horizontal at the top of the axis is much better than one parallel to the axis. Figure 10.2 shows examples. The only difficulty is that a long title must be presented on more than one line.

Axis labels (and all text on graphs and drawings) are best in a sanserif font such as Helvetica or Arial (end of Chapter 5). Such fonts are designed for short labels, while serif fonts such as Times New Roman are designed for readability in extended text.

Finally, on Tufte principles, do not label more axes than necessary. For example, where you have two or more parallel graphs in the same figure, with the same axes, you can omit one axis title, as in Figures 9.5 and 8.4.

Data symbols

Always show the data points, not just a line fitted to them.

When possible, choose the symbol to reflect the variable. For example, use a filled circle for the shaded plants, and an unfilled circle for the unshaded ones, or filled for experimental and unfilled for 'control' or reference. Treatments 1 to 7 may be matched with the number of sides in the symbols 'Point' (a small filled circle), upright or diagonal cross, triangle, square or diamond, pentagon, hexagon, and circle.

Two different sizes of the same symbol can be used in a way similar to filled and unfilled ones.

Keep the same symbols for the same thing or treatment throughout an article.

If numerous data points overlap then prefer unfilled symbols (the overlap is more likely to be distinguishable: Figure 8.5). An alternative is to add the number of superimposed points as an adjacent annotation.

If you use colour then recall that the figure may be photocopied in 'black on white' and choose the colours with sufficient contrast in density that they have clearly different shades of grey in the photocopy (Figure 8.1).

Lines

Fitted lines should not be plotted outside the range of the data (on the independent axis). Extrapolation is dangerous because you do not know whether the same trend continues or not. In the mid twentieth century it was assumed for decades that the line fitted to the graph of genetic damage against dose of ionising radiation could be extrapolated from the smallest measured dose back to zero at which it was pointing, with the conclusion that there is no threshold to genetic damage. Only

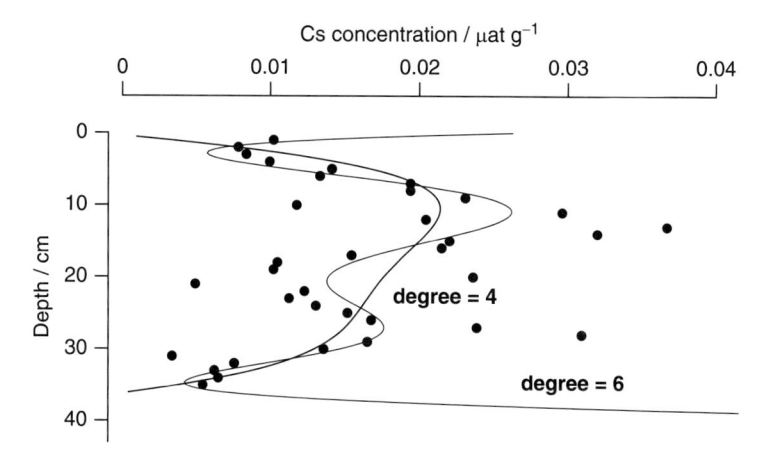

FIGURE 10.4 Polynomials fitted to the upper three-quarters of the data as in Figure 10.5. Thicker line: 4th degree (quartic); thinner line 6th degree (sextic). The quartic is the smallest degree that can reveal two peaks, but they are unclear. Extrapolating below (or above) the used range gives results that we know (from data between 35 cm and 45 cm deep in Figure 10.5) are grossly wrong. The sextic polynomial shows the peaks better but is even more misleading when extrapolated only a small way outside the data range. Compare this Figure to the 'LOESS' and smoothed patterns in Figure 10.5. Which Figure do you think is the more helpful?

later when more sensitive measures of damage were devised was this found to be untrue.

It is specially dangerous to extrapolate a fitted polynomial, the danger increasing with the degree of the polynomial (Figure 10.4). The fit has been got by a nice balance of linear, quadratic, cubic, quartic ... terms within the measured range. Even a small distance outside that range the line is likely to swing violently away from the trend. Ten data points could be fitted almost exactly by a ninth-degree polynomial, but it would not be sensible to do this. Polynomials may be useful to show large-scale trends, but they rarely carry understanding with them. It is unusual to find that degrees beyond three, or four at most, are useful.

Where you have lines connecting points directly, or adjacent bars in a bar chart, then the occasional erratic point will have long and conspicuous lines that can distract attention from the general

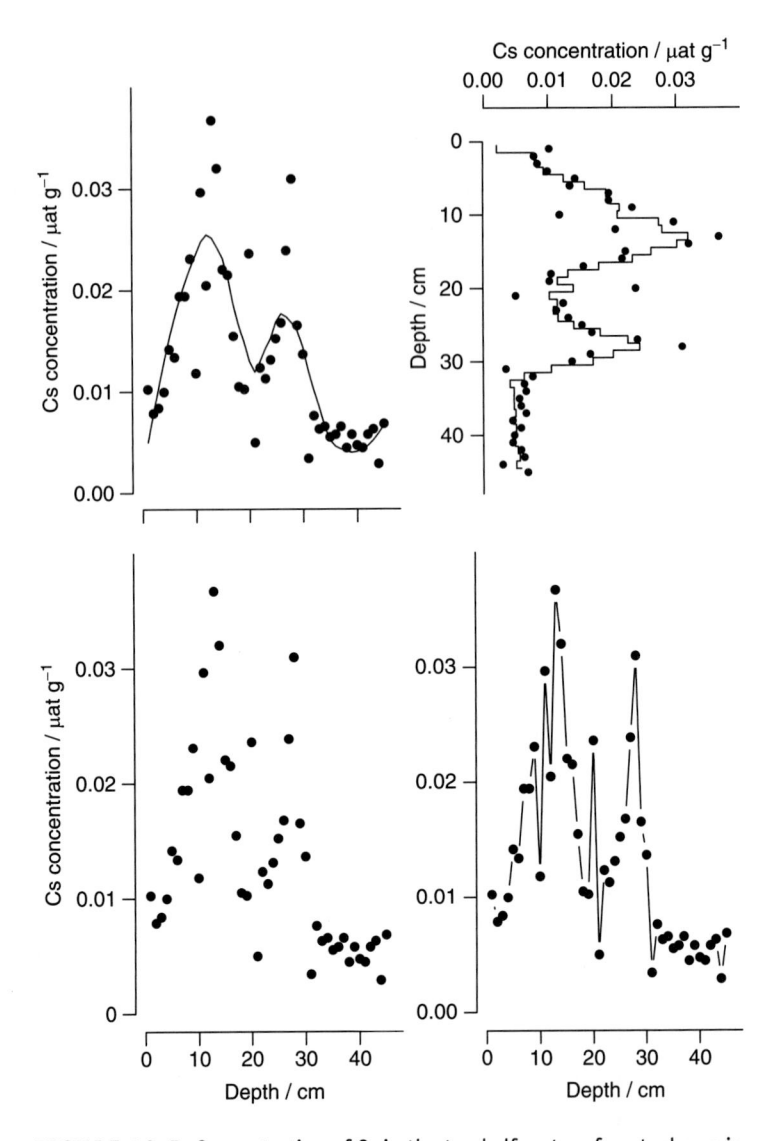

FIGURE 10.5 Concentration of Cs in the top half metre of peat, shown in four ways. Samples of about 10 cm³ were taken (without homogenising, for reasons not relevant here) from slices of 325 cm³. This results in occasional erratic values. *Lower left*: Just the data. Some indication of two peaks but not at all clear. *Bottom right*: Lines joining the points added. The longest, and most conspicuous, lines are to and from erratic points. This distracts attention from the overall trends. *Upper left*: Line for locally weighted regression (LOESS) added. Two peaks are clear. *Upper right*: axes swapped so the abscissa is the dependent concentration and depth is in its natural orientation and direction as the ordinate, and smoothed values added. Two peaks are clear again. Both LOESS and smoothing require choices that enhance or damp overall trends, and to that extent are not objective.

shape. An example is shown in Figure 10.5, lower right. The bare data and data connected by lines are in the lower half. Two ways of revealing the trend that is obscured by variation and long lines are in the upper half. 'Locally weighted regression' ('LOESS', 'LOWESS') at the upper left gives a continuous line the position of which responds most to the nearest point and decreasingly to more distant points. At the upper right are smoothed values for specific points, got by a calculation (Velleman & Hoaglin 1981) that is, again, influenced most by local points. Both show that there are two clear 'humps', the upper larger than the lower. Both methods can be 'tuned' to change the strength of response to local values and thus to emphasize or diminish local variation. To that extent, both methods are subjective and should be used with caution. The LOESS method gives a continuous line and so can be used more easily for interpolation.

Five kinds of line 'whiskers' are commonly used to indicate dispersion (Chapter 8 and Figure 8.5). One-sided whiskers give the 'exploder plunger' effect in bar charts (but for quantiles, both should always be shown because they are usually of different lengths). The descriptive spread of values about a central result is shown by (1) range, (2) SD, or (3) by quantiles (commonly quartiles or normiles) that may be asymmetric. A quick indication of the significance of differences can be got from (4) mean $\pm$ 2 $\times$ SEM or from (5) 95% CI (confidence interval) whiskers. If n is less than 25 or so CI is much to be preferred to SEM for the reasons given in Chapter 8. It is essential to name the sort of whisker(s) and, if CI, the span of the CI (which is not always 95%).

Annotations

In the days when figures were cast in lead alloy and mounted on wooden blocks, the blockmaker made corrections by cutting out pieces of the metal and nailing in replacement pieces. That was expensive, so editors encouraged authors to put as much of the (error-prone) writing as possible in the caption. That meant that the reader had to work out what the different curves or bars represented – in effect to learn a simple language or code for each graph. The data symbols were either described ('triangles') requiring translation into shapes, or shown by crude symbols that were rarely identical to those on the figure itself. Much cheaper processes are now used, and there is little cost incentive to minimise annotations on the graph. You can identify point and line types in a box (often confusingly called a 'caption' in graph-plotting programs), but this still requires the reader to cross-refer from the symbol in the 'caption' to the graph itself – to learn a (different) simple language. It is usually better to identify each curve by text, usually one word, on the graph (examples in Figure 4.2 and 4.4), and with an arrow if necessary. The text can be a cue, an *aide mémoire*, to a fuller description in the text caption below the figure.

Types of graph

The types of graph are uncountable – scan the website of any of the graph-drawing computer applications. Many (for example triangular graphs to display proportions of three components in a mixture) have been devised for special or even unique situations, and it is impossible to consider here most of these less common types of graph (for example three-dimensional, '3D', projections on to two dimensions, phase diagrams). Two-dimensional, '2D', scatter plot and bar graphs are the commonest in scientific articles. The sections above are relevant to simple '2D' scatter plots. Next, consider bar graphs of various kinds.

Bar graphs (bar charts)

Tufte principles would require only a small mark at the necessary position on the data space: all else is 'chart junk'. But in practice scientists are so used to bars (usually vertical, and for simplicity I assume that in what follows) that they would be distracted by a pure Tuftean graph. Nor would it be possible to distinguish the three sorts of bar graph: categoric, integer, and continuous (most often a 'frequency' histogram). These types are compared in Figure 10.6.

291　FIGURES: GRAPHS

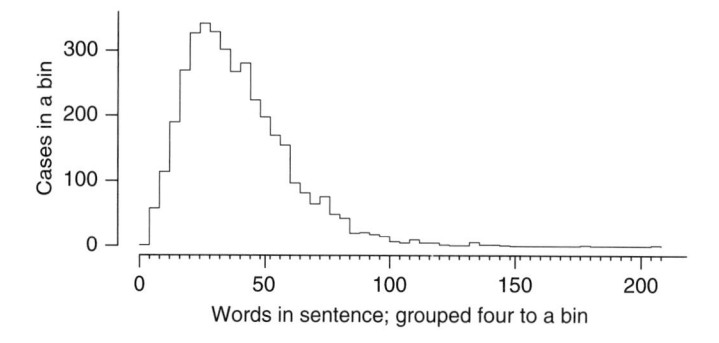

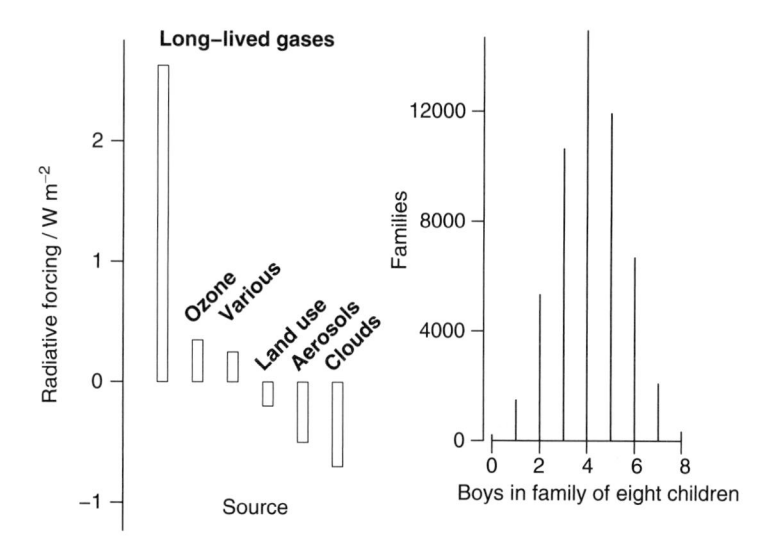

FIGURE 10.6 Three types of bar graph. *Lower left*: categoric radiative forcing factors (IPCC 2007) for long-lived gases (CO$_2$ 1.66, CH$_4$ 0.47, halocarbons 0.34, N$_2$O 0.16), tropospheric ozone, five smaller sources, and two negative sources: five aerosols, and reflection from clouds. *Lower right*: discrete (integer classes) of the number of male children in 53,680 families of eight children. *Upper*: 'continuous' (actually groups of four integers) histogram of frequency of length of sentences in Darwin's 'Origin of Species'. The same data as Figure 8.3 but in classes (bins) of four.

- Categoric bar graphs

    Each bar has noticeable width and the bars are separated by a noticeable space. The bars can be ordered by: (1) the natural order of their category (for example 'Previous', 'Current', 'Next') or

(2) if there is no natural order, by the height of the bar, or (3) if the reader will want to locate a particular bar at once, by alphabetic order (for example examination score by person's name or identity number).

- **Discrete (integer) bar graph**

    Each bar is a line (of a near-minimum visible thickness) placed at an integer value on the horizontal axis. The number of boys in families of eight children is an example.

- **Continuous ('step') bar graph**

    Vertical lines are minimised to 'stair risers' that connect the end of one horizontal bar to the start of its neighbour (plus, sometimes, at the left of the first bar and the right of the last, and on the preceding and succeeding bars to one that has zero numerical value) and the only gaps between bars are for zero or missing numerical values. The graph appears as a set of steps. The characteristic of this type is that the horizontal axis is continuous and is divided into 'classes', usually of equal width. Deciding how many classes to make requires judgement, and depends on how the data are distributed. To get the class width, an initial rule-of-thumb is to try dividing the data range into $\sqrt{n}$ equal classes. Suppose you have 89 measurements with numerical values in the range from 3.9 to 17.6. The first guess would be class width $(17.6 - 3.9) / \sqrt{89} = 1.45$. Then round that (up or down) to a user-friendly value, perhaps 1.5 in this example. But if we begin at 3.9 we get class boundaries at 3.9, 5.4, 6.9 ... So make the first (and last) classes of slightly different width to the others so the boundaries fall at 3.9, 5.5, 7.0, 8.5 and so on, so that all but the first and last boundaries are at user-friendly values, and all but the first and last classes have an easily understood width. An alternative is to make the classes of whatever fixed width you like and put ticks on the axis only at user-friendly distances, which now do not coincide with class boundaries. Which is the better method will depend on the nature of the data: is it important that the reader can see the value of the class boundaries accurately or not?

    This does reveal an important feature of such graphs: it is the width × height i.e. the *area* under a step that is important.

The vertical axis should be labelled as 'Cases per class' or, if the classes are all of the same width, 'Cases per class of width 1.5' or something similar. The calculated 'cases / width' must be made and used to place the step at the correct height.

If the graph is strongly skewed with only isolated instances to the right (or left) then it may be useful to group several of them into a wider step spanning several of the class widths used toward the left (or right), and of course the calculation of ' ... per class' must also be made.

The term 'histogram' (from the Greek 'histos', mast) is used if the vertical axis is a count or frequency or proportion or percentage, but is often used loosely for any kind of bar graph.

Finally, be careful with 'stacked bars' in which components are piled cumulatively on one another in a single bar. The overall height is easily judged, but comparisons of components, other than the basal one, are difficult because they begin at different heights.

'Pie' charts

Plain pie charts are widely used in commerce but in scientific reports should usually be replaced by bar charts. It is easier to compare a set of lengths than a set of areas (or the associated sector angles). In pie charts the sectors must be identified, commonly by shading or colour, and a separate caption. These patterns and especially colours (and colours photocopied as greys even more so) are hard to identify, while bars in a bar chart can have a text label next to (or in) them.

Some pie charts have sectors that extend out to different extents and can thus be used to show a second variable, but these too can be replaced by a more easily grasped rectangular graph.

Other radial charts

Some sorts of data naturally reflect the compass or clock. 'Wind rose' diagrams, in which the average wind speed or frequency from (often eight major) directions is shown by the length of radial lines, are easily understood. So are graphs showing the direction of a batsman's strokes in cricket, and clock faces showing different activities of an animal.

A similar principle can be used, for example, by representing the concentration of, say, seven ions in water samples by seven equally spaced radial lengths the tips of which are then joined. The general shape makes comparison of a lot of samples easy.

But we are now starting on an indefinite field for ingenious designs. The common feature is that relationships among complex data should be easily grasped visually.

Figures: other line diagrams

On geographic maps, always incorporate a scale and (compass) direction.

On line drawings always incorporate a scale bar. The old system by which a 'magnification' of '$\times 3$', or 'scale $\times 10\ 000$', or 'scale 1:10 000', or something similar was placed in the caption (or even in the drawing) ignored the fact that the drawing may be projected on a screen or reproduced on paper at a different magnification, so the written magnification is wrong.

Figures: images

Pictures – images – of blots, and gels, and of cell contents and ultrastructure are primary evidence in, for example, molecular biology and cell biology. Nowadays these images are mostly digital and they raise four technical questions: what colour mode should one use? what file format? what resolution? what manipulations are acceptable? These problems are fairly new and affect some journals more than others. Your target journal may make clear recommendations but if it makes none then general advice about colour modes, resolution, and file formats is in the Appendix to this chapter, while what follows concerns manipulations.

What image manipulations are acceptable?

Before digital images became usual few scientists had the darkroom expertise, or time, to do much more than vary the exposure of a print, or use a filter to enhance or diminish the visibility of structures and features selectively over the whole image, or to cut and paste images physically into a single figure. Digital images, however, can be modified easily by anyone with a small amount of skill. Such manipulations are widely used to 'clarify' a result, especially in digital micrographs, and in blots and gels (Rossner & Yamada 2004). These manipulations are often innocent (though mistaken), but some verge on (or step over into) deliberate misrepresentation. Here are a few suggestions about where the limits of 'acceptable' lie. (1) Use the settings on your digital recording apparatus (contrast, intensity range, colour) to achieve as nearly as possible the image that is to be published: you get better results by adjustments at this stage than by later manipulations. But avoid settings that are themselves so selective as to amount to misrepresenting the image. (2) Allow only cropping and manipulations that affect the whole image. If direct comparison with another image (for example a 'control') is to be made then apply the same manipulations to both images. No *individual* structure (for example one or a few cells, or a band on a gel) or feature should be enhanced, obscured, moved, added ('pasted') or removed. (3) Allow but report any global increase or reduction in brightness, or of colour in a limited part of the spectrum. But global changes of this kind must not remove structures or features (for example a faint band in a gel) from the image. (4) Do not use 'touch up' tools of any kind. (5) If you want to merge two images then leave a visible gap or line between them so that viewers realise what you have done. (6) Whatever manipulations you apply, keep a copy of the original, unmanipulated, image. It may be called for by the editor or a referee or (after publication) by a reader. (7) List in Methods or figure caption the device(s) used and significant settings, the resolution and size of the original image, and name any image-processing programs and what they were used for.

Chromosome image manipulations present special dangers. The original evidence is often an image of a metaphase plate with the condensed chromosomes spread about but usually with some overlapping. The convention is to 'isolate' ('cut out') each paired chromosome using digital manipulations and to parade them for discussion in an image in which the cut-outs are assembled in length order. These manipulations are open to mistakes and 'enhancements'. Good practice is to publish the original unchanged image of the metaphase plate without any manipulation as well as the parade. The practice of 'cleaning up' the background in such metaphase images is strongly deprecated. Some workers also leave gaps in the chromosome arms where chromosomes overlapped and one cannot tell whether one is seeing chromosome A or B.

QUESTIONS:
- For what sorts of data would you prefer a table, and for what sorts a figure?
- Why do most scientists use many more figures than tables?

Appendix to Chapter 10: colour modes, resolution, and file formats

Colour modes

Colour is expensive to print but on a computer screen costs no more than black on white. Screen and print use different modes to represent colour. What follows is but an outline of a complex subject.

Computer screens *emit* light from a rectangular pattern of pixels (dots) so close together that the dots are not distinguishable. Each pixel produces a desired colour by adding various proportions of red, green, and blue light. They create images in RGB mode (Red, Green, Blue) which is therefore 'additive'. Each component is often given as a value from 0–255 (needing 8 bits = 1 byte) so the full specification for a pixel needs 3 bytes = 24 bits and this can represent 256^3 = 16.8 million colours. (A fourth pixel may encode transparency.) If all three colours are at their minimum (0) then the screen appears black; if all are at their maximum (255) the screen appears white. If all three components have the same value the screen appears as one of 256 densities of grey, ranging from black to white. A greyscale image needs only 1-byte pixels (each used to code all three of R, G, and B). A line diagram needs only 1-bit pixels (either 0 or 1) i.e. 8 pixels per byte. Table 10.6 gives the RGB composition of some basic colours. Most scanners, digital cameras, and video machines create images in RGB (or a similar) mode in the expectation that they will most probably be used with a light-emitting device. The same effect as RGB is produced by HSV (Hue,

TABLE 10.6 Composition in the RGB mode of some colours. Note the extra line space separating white, blank, and mid-grey from the other colours

Name	Red	Green	Blue
Red	255	0	0
Green	0	255	0
Blue	0	0	255
Yellow	255	255	0
Cyan	0	255	255
Magenta	255	0	255
White	255	255	255
Black	0	0	0
Mid-grey	128	128	128

Saturation, Value) mode, though these terms are less intuitively obvious than Red, Green, and Blue.

Print relies on pigments that absorb light selectively and *reflect* what has not been absorbed. Images for printing are usually needed in CMYK mode (Cyan, Magenta, Yellow, blacK), which usually needs 4 bytes = 32 bits to specify the colour of a single pixel. C, M, and Y act as filters, selectively absorbing, and thus removing light from what falls on the paper. Thus CMYK is said to be a 'subtractive' mode. Cyan is the complement of red, absorbing red and reflecting green and blue light; similarly magenta is the complement of green, and yellow of blue. If C, M, and Y are at their minima the paper appears near white; at their maxima, however, because of imperfections, the paper appears a dark muddy brown, not pure black. The blacK component is therefore used to correct this, and for black text. As with RGB, equal contributions from C, M, and Y produce grey, and a part of the full colour can then be transferred to the black component, thus saving on the CMY colours. The CMYK colour space is smaller than the RGB one and some colours seen in RGB cannot be produced exactly in CMYK. The appearance on paper is strongly affected by the type and absorbance of the paper (compare a newspaper and a glossy magazine), by the specific inks used, and by the spectral composition of the illuminating light. Thus CMYK 'white' may appear bluish if the illuminating light is bluish. For these, and other, reasons CMYK images often look duller than the equivalent RGB ones and can appear to be substantially different from what you expected. Producing an accurate reproduction on paper of the colour you saw in the microscope is an art, depending as much on the skill of the printer and on the light used for viewing as on anything you can do.

Many journals still consider their printed version as their 'primary' record, with the electronic version as subsidiary. They therefore need CMYK images for printing and prefer, even if they do not require, this mode. If you supply an RGB-mode file the journal will convert it to CMYK, but the process is complicated and there is no single method for doing it. The result may then surprise and disappoint you. Some journals, however, now treat the electronic file as the primary record and the printed version as derivative. They prefer or require digital images to be in RGB mode and convert, with some distortion of colour, to CMYK for their printed version.

Colours for the 'colour-blind'?

Among 100 readers or viewers you may expect about 8 to have some defect in colour vision. The commonest affliction is in men who are unable to distinguish red from green. If you have an image of a microscope slide with red(dish) and green(ish) double staining then a significant number of your viewers will be unable to see any distinction. What can you do?

In a slide for presentation, just avoid red and green in the same slide if you can. If not, try transforming red to purple (for example).

In a colour print, try substituting red in the image with magenta (and declare in the caption that you have done so). If the image is coloured but colour printing is not to be used, try greyscale. Differences in brightness may be visible. If not then try substituting one of the colours by a different one of distinctly different brightness and then print as greyscale.

File formats and image resolution

There are hundreds of file formats. To submit an article to a journal you probably need to know about half a dozen. Some are for text alone, some for pictures (images), and some include both. Most files are identifiable by their suffix (the letters after the last '.' in the name). All but simple text files also contain internal codes that identify the format, but you won't normally see these.

Simple text
Usual suffix '.txt', '.asc'. Each byte (or two bytes if extended Unicode is used) represents a character, usually in ASCII (American Standard Code for Information Interchange). Operating systems differ in how they mark the end of a line, but a file using one convention can be converted to another.

Word-processor
There are numerous word-processors whose native file formats are often incompatible, though most programs try to read the commonest formats (such as '.rtf', '.doc', and '.odt') and to write these and, especially, '.pdf' files (see 'Document layout' below). The word-processor programs can import, show, and incorporate images (as well as text and tables). Many journals require the final text of an article to be in '.doc' format. *Never* include images in the final submission: extracting an image from a '.doc' file for separate processing is problematic. Many journals require each image to be submitted in a separate file.

Document layout
Files with the suffix '.ps', '.pdf', '.tex' specify the layout and appearance of a document, as well as containing the content to be shown.

- ■ '.ps': PostScript is a language and a '.ps' file contains a program with commands and the data (for example as text, bitmaps, or '.eps' files) on which the commands are to operate. When the file is opened the program runs and thus creates the document.

A consequence is that you cannot go direct to, say, page 5 because that does not exist until pages 1 to 4 have been created.

- '.pdf' (Portable Document Format): unlike '.ps' files the pages exist as locatable units so one can jump to any page direct. This is a widely used file format for the *initial* submission of an article, and is required by many journals.
- '.tex': LaTeX, like PostScript, is a language. A '.tex' file contains commands and the content they are to operate on. LaTeX handles the details of high-quality typesetting so that the user can concentrate on the content. LaTex might be considered a word-processor too. Some scientific communities (for example, physics) use LaTeX routinely, but many bioscience journals will not accept articles in this format.

Images: general

Images may be as a 'rastered bitmap' (a series of dots) or 'vectored' (instructions to draw lines from here to there). All figures must eventually be converted if necessary by the printer to a bitmap format for printing on paper or displaying electronically.

For bitmaps, publishers specify the minimum resolution your images must be at. For example they may specify at least 300 dpi (dots per inch) for colour and greyscale images, and 1200 dpi for line drawings including many graphs. It is up to you to specify to your plotting program the resolution and the height and width of the image, in pixels ('px'), mm, cm, or inches. Thus, for an image to be printed at 300 dpi at a scale of 5 inches high by 8 inches wide you might specify 'units = "px", res = 300, height = 5 * 300, width = 8 * 300'. Details of how to do this are program-dependent. You may be able to specify only one of height and width plus the aspect ratio (width / height), or you may specify only the width and the program then calculates the height from the data to be plotted. One way or another, all three values must be specified somehow.

Some publishers will also accept images as '.eps' files. These contain a vectored image of effectively infinite resolution whose height and width are specified (by you), and which the publisher will convert to a bitmap, choosing a suitable resolution.

A vectored image can be converted to a rastered one, and you can usually interconvert rastered images in one format to another rastered one. It is not generally possible to convert a rastered image to a vectored one.

Be aware that images on the web or for a presentation program are usually rastered and of lower resolution (72 dpi) than is needed for printing. These files are small and efficiently compressed in formats such as '.gif' or '.jpg' (see below) which load and display fast. If you use an image-manipulation program to increase the resolution (by calculated interpolation between existing dots) at the same image size from 72 dpi to 300 dpi then 95% of the dots have been produced by you, not by Nature.

Instead of coding a rectangle of dots a vector format records a series of lines drawn from one pair of coordinates to another pair (the method used by the most commonly used text fonts too). Graph-drawing or line-drawing programs may offer vectored file formats such as '.svg' (scalable vector graphics), and '.eps' (encapsulated postscript). Image files of these kinds are printable on their own, and may be embedded in, for example, a '.doc' file.

Bitmap (pixmap) image files
Suffices '.raw', '.gif', '.png', '.tif' (or '.tiff'), '.jpg' (or '.jpeg') and uncounted others. Most bitmap files can be compressed (so the file is smaller) and later decompressed without losing any information. The file will contain internal codes identifying the type of compression and the program that opens the file will automatically decompress it. This is 'lossless' compression. But '.jpg' and '.jpeg' files lose some information every time they are saved. If you open a '.jpg' file and close it several times in succession the image becomes more and more fuzzy. Here are some details about the formats.

The bitmap for a simple black on white graph consists of a series of '0s' and '1s', the '0s' representing the white background and '1s' the black foreground. An illustrative part of the file might contain '000000000...000111111100000', the '...' representing numerous more '0s' and the '111111' where the scan crosses a line. The compressed version might be '2717 '0s', 6 '1s', 4762 '0s'' and so on (though more efficiently stored than this). If the graph is a simple one then the saving in space can be enormous. A 5" × 6" black-on-white graph creates an uncompressed file of about 40 MB which, when compressed, occupies only 20 kB: a 2000-fold reduction in size. Always compress files (such as '.tif' files – see below) that allow it.

- '.raw': the unprocessed format for digital camera (and other) devices. This format produces very large files with no manipulations or compression. There is no standard – each manufacturer has its own, though many are moderately similar. Before publication you must convert to a format the publisher can handle. The advantage of '.raw' format is that the file contains all that the equipment that made it is able to record, and is therefore the best base for later processing.
- '.gif': an old format with each pixel occupying a single byte. This can be greyscale or colour 'indexed' by numerical value to one of a restricted palette of 256 colours which can be from a standard palette or can be automatically adapted to give the best results for the particular image encoded in the file. Compression using the LZW method is lossless. This format is little used for photographs, but it may still be useful for diagrams (and for the web).
- '.png': a replacement for '.gif'. The standard allows RGB, grey, and line pixels, indexing and compression. It can be used for almost any image, and is common on the web, but is not what printers need for coloured images because it uses only the RGB colour mode, not CMYK.
- '.tif', '.tiff' (tagged image file format): there is a basic standard that should be understandable by any program claiming to accept '.tif' format, but the design permits other sorts of information to be included, each introduced by an identifying 'tag'. This licence has been widely used for non-standard additions. The standard allows RGB, grey, and line pixels and indexing as in '.png'. Lossless compression can be by LZW or ZIP methods. The standard also caters for other colour modes, particularly CMYK, and this makes '.tif' the almost universal choice of printers (the people) and

therefore for coloured figures accompanying an article submitted for publication.

- '.jpg', '.jpeg' (Joint Photographic Experts Group): RGB colour only. Is automatically compressed but the decompressed image will have lost some of its details permanently. The severity of compression can be specified, the greater the compression the greater the loss of detail. Losses are cumulative. Every time you open and then save a '.jpg' file you lose something even if you set the compression 'loss' to '0'. For this reason '.jpg' files are unsuitable for primary evidence (such as, for example, photographs of gels, microscope images) but can be acceptable sources for secondary purposes such as photographs of apparatus or of an ecologist's field site. To avoid loss of detail, any '.jpg' file intended for publication should be immediately converted to '.tif' format (or, better, recorded in '.tif' or '.raw' format in the first place).

Is there a simple strategy to guide you through this minefield? Yes. See what your target journal requires! If that fails, then use '.eps' for line images you make yourself, and '.tif' or '.tiff' (or, less good, '.jpg' or '.jpeg') for images made by devices (cameras, scanners) that can do no better.

11

Citing and referencing

After the main text of your article there is a section, usually headed References, containing bibliographic details of those publications that you have used for support in the main text. Each entry contains information about items such as author(s), date of publication, title, name of publication, and pages, given in at least sufficient detail to allow a reader to locate the work. In the main text you point to a particular reference with a cue or key. The cue+reference is a citation, but the cue or key is itself commonly called a citation, and I follow that custom here. There are many styles of citation ▼ (cue) and reference, and each journal makes its own choice. In many journals the 'Instructions to Authors' are vague about the citation and reference system to be used, and about formatting details for both, but you can infer these from examples in recent issues of the journal or get a reference program to use the journal format template if it offers one.

> ▼ Citations: a suggestion. References in scientific work are conventionally given to whole articles or books. But the material relied on in a citation is often only a small, perhaps tiny, part of the whole reference. For example, I may cite Chason & Siegel (1986) for values of the hydraulic conductivity of peat. But a reader wishing to check my claim would be helped by knowing

> that the values I rely on are in Figure X on page Y. This could be accommodated by the citation: Chason & Siegel (1986, p. 95) or Chason & Siegel (1986, Figure 6). The reference needs no change and can be used for further citations to different specific parts. A search over 10 pages is not too time-consuming. But suppose I had cited Glasstone (1962) – the thickest textbook on the shelf beside me, with1320 pages. Although the Table of Contents is detailed and there is a good index one would spend some time in a search.
>
> The legal profession requires this sort of detailed citation. Perhaps scientists should follow their lead?

Citations

Author–date system

Here is an example of 'author–date' citations, sometimes called the Harvard system, in one style using space and ',' as separators; 'and' to link two authors; '*et al.*' for three or more authors; and multiple citations ordered by date:

> '... by Brown and White (1998b) contrasts with earlier views (Smith 1991, Jones *et al.* 1995)'.

The '1988b' implies citation somewhere of another article by Brown and White designated '1988a'.

Another, differently formatted, version of this citation is:

> '... by Brown & White (1998b) contrasts with earlier views (Jones, Williams & Evans, 1995; Smith, 1991)'

This style uses ',' and ';' as separators; '&' to link authors; naming all three authors (at least on first mention, but probably requiring '*et al.*' for later citations and probably always for four or more authors); and multiple citations ordered alphabetically by first author.

If your target journal uses the author–date system the editor will have chosen a particular set of these elements – separators, links, multiple author names, and multiple citation order – which you can infer (if the journal is shy about these details) by inspecting a recent example of the journal.

In the author–date system the references themselves are arranged in alphabetical order of first author, which makes it easy to find a cited reference. But there is no easy way to discover quickly where in the text of a printed article an interesting reference has been cited. (A search of an electronic text is easy though.)

The author–date citation system is easy for you, as author, to use. If at a late stage you want to add a citation to a new reference you simply insert the reference in its correct alphabetic place. But the citations themselves take non-trivial space in the text, and readers probably want to skip most of them, on first reading at least. Strings of citations, most likely in the Introduction and Discussion, can be a significant distraction for readers.

Citation-number and reference-number system

For these, and other, reasons some journals require citations as numbers – a system that is less immediately informative than author–date but takes less space in the text and causes a smaller distraction for readers. The numbers are usually superscripts or are enclosed in brackets (or sometimes in parentheses):

'... new views[17] contrast with earlier ones[32, 27] ...'

or, in bracketed form,

'... new views [17] contrast with earlier ones [32, 27] ...'

There are two variants of this system, depending on the order in which numbers are allocated. In the 'citation-number' system numbers are used consecutively from the start of the text; in the 'reference-number' system references are arranged in alphabetical order, then numbers allocated consecutively to them. In the examples above one can infer that either the 'reference-number' system is being used, or that there have been earlier citations of [17] and [27], and that we have just allocated [32].

At times you may wish to name one or more authors to indicate the authority for the citation. This is easily done: 'Bloggs[15,17] argues that ...'.

In the 'citation-number' system the references are arranged in order of citation, not alphabetically. This makes it difficult for a reader to locate in a printed article a reference whose author is known, but fairly easy to find where a reference was first cited. (Again, an electronic search is easy.)

Both 'reference-number' and 'citation-number' systems are awkward for you as author because any late addition or removal of a source is likely to require wholesale changes of all citations with a higher number: adding 'Aaby' to the start of the Introduction at a late stage could be seriously disruptive. In practice a fudge may be used. An inserted reference may be cited as, for example, '[27a]'. A better solution is to use a reference-managing program and let it make all the necessary changes.

A web search for 'Reference manager comparison' will reveal at least 25 such managers. Some are able to interact with common word-processors and insert citations and references into your article in the correct format for your target journal.

References

A Bibliography is a list of works that you have consulted and which one can assume have influenced your thinking in some general way. Scientific reports rarely have a Bibliography. If you want to point to a work in this background rôle you can cite it, and give details as a reference. References are sources that you have specifically cited in the text.

The three commonest sorts of reference are to an article in a periodical ('journal'), to a chapter with one or more named authors in an edited book, and to a whole book with named authors or editors. Try to avoid citing the 'grey literature' – for example, internal reports, abstracts for meetings, and theses. Many universities now require that the bulk of a science thesis should be already published or accepted for publication, so there is no need to cite the thesis. But some theses contain useful material that has not, and may never be, formally published.

Make website references only if your target journal allows them, and then only to archives with some prospect of long-term existence. A recent investigation found that during the previous ten years, 20% of website references could no longer be found. The lost sites included one that contained an article about the unreliability of such references. If you *must* refer to this sort of source, or to an online journal (particularly if it has no paper version), then give the date on which you read it too: '*Journal of Wild Hopes* (online 27 Feb. 2012)'. And keep a paper copy.

Websites often change, even if they have not disappeared, and your readers may discover that what you rely on has been removed or, worse, changed without any indication of this.

Here are the minimal requirements. For a cited article in a journal you must give the author(s), year of publication, article title, journal title, the volume number and page numbers (first, and usually last). For a book you must give author(s) or editor(s), date, title, publisher and place of publication. For a chapter in an edited volume you must give the author(s), date, chapter title, and in addition name the editors and give the volume title, publisher, and place of publication too.

Journals differ irritatingly in the details they expect you to present, and the order and format in which to present them. Getting the references right is tedious but necessary. It is your responsibility to do this, not the editor's. Choosing the target journal before starting to write helps considerably. So does using one of the reference-managing computer programs that insert citations into an article and, when instructed, compose the reference list in the format you specify. These programs nowadays have numerous templates for named journals that will apply the necessary format automatically, or the journal itself may supply the necessary template.

Authors are always identified primarily by their family name, then by the initials of their given name(s). It may not be obvious in, for example, transliterated Chinese or in Spanish names, which of the elements is the family name.

Some works become known by their author's name, and that name is kept when the work is revised by a different person. Thus in 'Fowler revised Gowers (1968)' Fowler was the original author and Gowers the reviser. In 'Gowers revised Greenbaum & Whitcut 1986' Gowers was the original author. His work was first revised by Fraser, and later by Greenbaum and Whitcut. By convention Fraser's part in the process is now lost: cite and reference only the latest reviser(s). Henry Gray's *Anatomy of the Human Body* reached its 40th edition in 2008, leaving, like an aircraft contrail in the sky, an evaporating cloud of earlier editors and contributors.

Some journals omit article titles. Readers cannot then assess easily whether the article is worth investigating or not.

An increasing number of journals also quote the Digital Object Identifier (DOI), if one exists, after all the other information, or even in preference to it. An example is *doi*:10.1029/2002GB001983. The DOI identifies a particular piece of intellectual property. The '10' identifies this as a DOI, and the '1029' is the organisation (the American Geophysical Union in this example) that is registered to allocate DOIs. Together these numbers before the '/' are the prefix and are allocated by the central DOI agency. What follows the '/' is allocated by the registered organisation, and must be unique within that organisation. It can contain useful information (the GB is for the journal Global Biogeochemical Cycles, and 2002 is the publication year) but in general it need not. If you type in a web browser, <www.doi.org> (6 Sept. 2013) and when you reach the webpage enter the full DOI, you get the 'metadata' associated with the DOI. This particular example gives bibliographic information, the article Summary, links to the sponsoring organisation, and instructions about how to get a copy. In general the metadata are bibliographic, commercial, and address or location data. A small problem is that sometimes the year in a DOI differs from the year of publication of the journal.

In the author–date and reference-number systems the references are arranged in alphabetical order. References to articles by the same first author (for example, Jones) are usually put into three groups. First are articles by Jones alone, in date order from earliest to most recent. Second are articles by Jones and one other author, in alphabetic order of second author, and then by date for articles by the same two authors. Third are articles by Jones and two or more other authors. Sometimes these are ordered alphabetically on second, then third, then fourth author, and so on. Some journals simply use date order for this group though. The settings in a reference-manager program will show how numerous are the possibilities.

Recall that references, like caryatids, ▼ are for support not just for decoration. Nor are they to boost the citation indices of colleagues. You should have read the work you cite and formed your own opinion about it, not just copied the reference or opinion from someone else or a database. Many widely cited works say something quite different from what their citers claim. •

> ▼ Caryatids are carved stone statues of draped female figures that act as pillars. Examples are the six that support the roof of the porch of the Erechtheion on the Acropolis in Athens.

> • Dobell (1938) in an obituary assassinates a ghostly author, O. Uplavici, who '... though [he was] a pure Czech had a Greek father and a German mother. He was born in 1887, published his only paper in the same year, obtained his doctor's degree later in the United States, and now – after a chequered career in many countries – breathes his last, as I write, in England.' What follows is from Dobell's account. The distinguished physician Jaroslav Hlava published in Czech – Hlava J (1887) Časopis lékařův českých 26(5) – an article about work in which he caused dysentery in cats by transferring to them material from human faeces probably containing *Entamoeba histolytica*. The article was titled 'O úplavici. Předběžné sdělni' which translates as: 'On dysentery. Preliminary Communication'. The work was

reviewed briefly in German – Kartulis S (1887) *Centralblatt für Bacteriologie und Parasitenkunde* 1:537 – but Hlava's name as author was omitted and substituted by Uplavici O. The reviewer, Kartulis, even writes of his correspondence with Uplavici. The Contents lists 'Uplavici, O'; while the Subject Index lists 'Hlava, Uplavici' and the Author Index lists only 'Hlava'. The later widely scattered literature has references to 'Hlava'; 'O.Hlava'; 'Hlava, Uplavici'; 'O. Hlava (O. Uplavici)' and even to 'Hlava and Uplavici'. The final accolade of an honorary doctorate was given by the *Index-Catalogue of Medical and Veterinary Zoology* which recognised 'Uplavici, O. [Dr.]'. It seems doubtful if any of those who referred to the article (which is in Czech and was not translated) had read it.

Most scientists who have tried to track an idea to its birthplace have had a similar experience.

Abbreviated titles of journals

Some journals do require abbreviated titles of journals; others do not (or allow the author to choose either full or abbreviated titles). Why abbreviate? The only good reason is to save space. If a full title makes the reference run onto another line, and if this happens so many times in a single list of references that another page is necessary, then space is saved by abbreviation. The saving may be illusory.

There are several lists of abbreviations. Some try to be general, others are for a specific field. Books of abbreviations are out of date by the time they are published, so it may be better to use an electronic list. But there are no generally agreed rules about how an abbreviation should be made, and lists differ in how they abbreviate many or most titles.

The makers of such a list should ensure that no two of their abbreviations are the same, but they are not much concerned with avoiding ambiguity for the user. The reader of an article is left wondering whether, for example, '*Res*' came from '*Results*' or

'Research'; whether 'Int' is contracted from 'Internal' or 'International'. This makes for difficulties when trying to locate a reference either electronically or through a library. It may therefore be best to make your own abbreviations if you are forced to abbreviate. Here are some guidelines. (1) Omit 'a' and 'the' and 'link' words such as 'of' and 'and'. (2) If the title is then a single word, do not abbreviate it. (3) Do not use a single-letter abbreviation except 'J.' for 'Journal' and 'Z.' for 'Zeitschrift'. (4) Do not abbreviate if to do so will save fewer than two letters (if you are adding a full stop to the abbreviation this means saving at least three characters). (5) Most words in journal titles can be put into

TABLE 11.1 Common words in journal titles and suggested abbreviations

Journal type		Journal subject[a]		Geography		Other qualifiers	
Advances	*Adv.*	Biochemical	*Biocheml*	American	*Amer.*	Academy	*Acad.*
Annals	*Ann.*	Chemical	*Cheml*	Australian	*Austral.*	Annual	*Annual*[c]
Archives	*Arch.*	Computer	*Compr*	British	*Brit.*	Applied	*Appl.*
Bulletin	*Bull.*	Ecological	*Ecoll*	Canadian	*Canad.*	Experimental	*Experl*
Comptes Rendus	*C. R.*	Engineering	*Engg*	Chinese	*Chin.*	Industrial	*Industl*
Journal	*J.*	Geological	*Geoll*	European	*Europ.*	Inorganic	*Inorg.*
Proceedings	*Proc.*	Hydrological	*Hydroll*	Russian	*Russ.*	Institute	*Inst.*
Progress	*Prog.*	Mathematical	*Mathl*			International	*Int.*
Reviews	*Rev.*	Marine	*Mar.*			Organic	*Org.*
Reports	*Rep.*	Microbiological	*Microbioll*			Quarterly	*Quart.*
Transactions	*Trans.*	Molecular	*Molec.*			Research	*Res.*
Zeitschrift	*Z.*	Oceanographical	*Oceanogl*			Scientific	*Sci.*
		Physical[b]	*Physical*				
		Physiological	*Physioll*				
		Zoological	*Zooll*				

[a] Most subjects appear in two forms, for example *Chemistry* (noun) and *Chemical* (adjective). The noun would be *Chem.* and the adjective *Cheml* without a full stop because the abbreviation ends in the same letter as the full word. Note the doubled 'l' where the noun abbreviation ends in 'l'. Some journals do not capitalise the initial letter of adjectives.
[b] Physics would be 'Phys.', but could be confused with Physiology, so use 'Physics'. The adjective 'Physical' is then unabbreviated as it saves one letter at most.
[c] 'Annual' is unabbreviated to avoid confusion with Annals, though one is a noun and the other an adjective.

one of four categories: Type, Subject, Geography, and Other qualifiers. Table 11.1 contains suggestions.

QUESTION:

What would you do if you had expanded an abbreviated journal reference but could not then find that name listed anywhere?

ENVOI

Here is the message again, for the last time: aim to write, speak, and illustrate clearly and concisely. Use conservative grammar, not to pander to the prejudices of grammarians and pedants but because you aim to avoid distractions for your readers, for the members of your audience, and for your visitors.

To communicate well is a useful skill. More than that though, it is a service to all those who work in hope in the anthill that is the city of science.

Bibliography and references

The style of entries here minimises punctuation, consistent with readability.

Bibliography

The following books and articles concern writing in general (shown by '*'), and particularly writing and presenting scientific work and reports. Some of these publications appear again as specific citations in the text. I found those with a '†' particularly helpful. This list is not comprehensive: 'Of making many books there is no end' (Ecclesiastes 12:12).

Alley M 2003 *The Craft of Scientific Presentations*. Berlin, Springer. 241 pp.

Barrass R 1978 *Scientists Must Write*. London, Chapman and Hall. 175 pp.

Booth V 1993 *Communicating in Science* 2nd Edn. Cambridge, Cambridge University Press. 78 pp.

Cargill M, O'Connor P 2009 *Writing Scientific Research Articles*. Chichester, Wiley-Blackwell. 173 pp.

Day RA 1998 *How to Write and Publish a Scientific Paper* 5th Edn. Westport, Connecticut, Oryx. 275 pp.

Dawson MM, Dawson BA, Overfield JA 2010 *Communication Skills for Biosciences*. Chichester, Wiley-Blackwell. 169 pp.

Divan A 2009 *Communication Skills for the Biosciences*. Oxford, Oxford University Press. 270 pp.

†Gopen G, Swan J 1990 The science of scientific writing. *American Scientist* **78**: 550–558

†*Gowers E, revised Greenbaum S, Whitcut J 1986 *The Complete Plain Words* 3rd Edn. London, Her Majesty's Stationery Office. 288 pp.

*Fowler HW, Fowler FG 1906 *The King's English* 2nd Edn. Oxford, Oxford University Press. 370 pp.

*Fowler HW, revised Gowers E 1968 *A Dictionary of Modern English Usage* 2nd Edn. Oxford, Clarendon Press. 725 pp.

Gustavii B 2003 *How to Write and Illustrate a Scientific Paper.* Cambridge, Cambridge University Press. 141 pp.

Hall GM (ed.) 2003 *How to Write a Paper* 3rd Edn. London, BMJ Publishing. 176 pp.

†Harmon JE, Gross AG 2011 *The Craft of Scientific Communication.* Chicago, Illinois, University of Chicago Press. 240 pp.

Johnson S, Scott J 2009 *Study and Communication Skills for the Biosciences.* Oxford, Oxford University Press. 235 pp.

†Kirkman J 1980 *Good Style for Scientific and Engineering Writing.* London, Pitman. 131 pp.

Matthews JR, Bowen JM, Matthews RW 2000 *Successful Scientific Writing* 2nd Edn. Cambridge, Cambridge University Press. 235 pp.

*Partridge E 1990 *Usage and Abusage* 3rd Edn. London, Penguin Books. 219 pp.

Peat J, Elliott E, Baur L, Keena V 2002 *Scientific Writing: Easy When You Know How.* London, BMJ Publishing. 292 pp.

*Strunk W Jr, White EB 2000 *The Elements of Style* 4th Edn. Boston, Massachusetts, Allyn and Bacon. 105 pp.

Turabian KL, revised Grossman J, Bennett A 1996 *A Manual for Writers of Term Papers, Theses, and Dissertations* 6th Edn. Chicago, Illinois, University of Chicago Press. 308 pp.

*Turk C, Kirkman J 1982 *Effective Writing.* London, Spon. 257 pp.

†*Williams JM 1990 *Style: Towards Clarity and Grace.* Chicago, Illinois, University of Chicago Press. 208 pp.

References

These are cited specifically in the text.

Aarnio P and 561 others 1989 Measurement of the mass and width of the Z^0-particle from multihadronic final states produced in e^+e^- annihilations. *Physics Letters B* 231: 539–547

Aad G and 2925 others 2008 The ATLAS experiment at the CERN Large Hadron Collider. *Journal of Instrumentation* 3: 1–407. *doi*:10.1088/1448-0221/3/08/SO8003

REFERENCES

Adkins CJ 1987 *An Introduction to Thermal Physics.* Cambridge, Cambridge University Press. 134 pp.

Akaike H 1973 Information theory and an extension of the maximum likelihood principle. In: *2nd International Symposium on Information Theory* 267–281. Ed Csaki F. Budapest, Akadémiai Kiadó

Alder K 2002 *The Measure of All Things.* London, Little, Brown. 466 pp.

Anonymous 2002 Editorial: Reflections on scientific fraud. *Nature, London* **419**: 417, 419–420

Anonymous 2008 Editorial: Working double blind. Should there be anonymity in peer review? *Nature, London* **451**: 605–606, correction 453: 711

Bayes TR 1763 An essay towards solving a problem in the doctrine of chances. *Philosophical Transactions of the Royal Society of London* **53**: 370–418

Berger VW, Ioannidis JPA 2004 The Decameron of poor research. *British Medical Journal* **329**: 1436–1440

Biagioli M 2003 Rights or rewards? In *Scientific Authorship* 253–279. Eds Biagioli M, Galison P. London, Routledge.

Bohannon J 2013 Who's afraid of peer review? *Science* **342**: 60–65. *doi:* 10.1126/science.342.6154.60

Booth V 1993 *Communicating in Science* 2nd Edn. Cambridge, Cambridge University Press. 78 pp.

Broad W, Wade N 1982 *Betrayers of the Truth.* London, Century Publishing. 256 pp.

British Standards Institution 2005 BS 5261-2: *Copy Preparation and Proof Correction. Part 2. Specification for Typographic Requirements, Marks for Copy Preparation and Proof Correction, Proofing Procedure.* London, British Standards Institution. 20 pp.

Buck CE, Cavanagh WG, Litton CD 1996 *Bayesian Approach to Interpreting Archaeological Data.* Chichester, Wiley. 382 pp.

Bulley M 2006 Clunky writing is proof positive of lazy thinking. *The Times Higher Education Supplement.* 2 June: 14

Bulmer MG 1965 *Principles of Statistics.* Edinburgh, Oliver and Boyd. 214 pp.

Chicago University Press 2010 *The Chicago Manual of Style: The Essential Guide for Writers, Editors, and Publishers* 16th Edn. Chicago, Illinois, Chicago University Press. 1026 pp.

Cleveland WS 1994 *The Elements of Graphing Data.* Summit, New Jersey, Hobart Press. 297 pp.

Cleveland WS, McGill R 1985 Graphical perception and graphical methods for analyzing scientific data. *Science* **229**: 828–833

Close F 1990 *Too Hot to Handle: The Story of the Race for Cold Fusion.* London, WH Allen. 376 pp.

Clymo RS 1995 Nutrients and limiting factors. *Hydrobiologia* 315: 15–24

Clymo RS 2012 How many digits in a mean are worth reporting? *ArXiv*:1301.0170

Clymo RS, Williams MMR 2012 Diffusion of gases dissolved in peat pore water. *Mires and Peat* 10(6): 1–10

Colquhoun D 2007 How to get good science. *Physiology News* 69: 12–14

Conway Morris S 1998 *The Crucible of Creation: The Burgess Shale and the Rise of Animals.* Oxford, Oxford University Press. 242 pp.

Copeland BJ (ed.) 2006 *Colossus.* Oxford, Oxford University Press. 462 pp.

Darwin C 1859 *On the Origin of Species by Means of Natural Selection or the Preservation of Favoured Races in the Struggle for Life.* London, John Murray. 498 pp.

Deichmann U, Müller-Hill B 1998 The fraud of Aberhalden's enzymes. *Nature, London* 393: 109–111

Delamothe T, Smith R 2004 OA publishing takes off. *British Medical Journal* 328: 1–3

Dobell C 1938 Dr. O. Uplavici (1887–1938). *Parasitology* 30: 239–241

Dyson F 2004 A meeting with Enrico Fermi. *Nature, London* 2004 **427**: 297. *doi*:10.1038/427297a; with permission from Macmillan Publishers, (c) 2004

Fang FC, Steen RG, Casadevall A 2012 Misconduct accounts for the majority of retracted scientific publications. *Proceedings of the National Academy of Sciences of the USA* **247**: 17028–17033. *doi*: 10.1073/pnas.1212247109

Farman JC, Gardiner BG, Shanklin JD 1985 Large losses of total ozone in Antarctica reveal seasonal ClO_x/NO_x. *Nature, London* 315: 207–210. *doi*: 10.1038/315207a0

Feynman RP, Leighton RB, Sands M 1963 Chapter 22 (10 pp): *Algebra. Lectures on Physics: Vol. 1: Mainly mechanics, radiation, and heat.* Reading, Massachusetts, Addison–Wesley. 535 pp.

Fowler HW, Fowler FG 1906 *The King's English* 2nd Edn. Oxford, Oxford University Press. 370 pp.

Fowler HW, revised Gowers E 1968 *A Dictionary of Modern English Usage* 2nd Edn. Oxford, Clarendon Press. 725 pp.

Frankel F 2002 *Envisioning Science: The Design and Craft of the Science Image.* Cambridge, Massachusetts, MIT Press. 335 pp.

Fugh-Berman AJ 2010 The haunting of medical Journals: how ghostwriting sold "HRT". *PloS (Public Library of Science) Medicine* 7: e1000335. *doi*:10.1371/Journal.pmed.1000335

Galison P 2003 The collective author. In *Scientific Authorship* 324–355. Eds Biagioli M, Galison P. New York & London, Routledge.

Gardner M 1981 *Science. Good, Bad and Bogus*. Oxford, Oxford University Press. 412 pp.

Garfield S 2010 *Just My Type*. London, Profile Books. 352 pp.

Godlee F, Jefferson T (eds) 1999 *Peer Review in Health Sciences*. London, BMJ Books. 271 pp.

Gordon JE 1968 *The New Science of Strong Materials*. London, Penguin Books. 269 pp.

Gould SJ 1989 *Wonderful Life*. New York & London, Norton. 347 pp.

Gowans CS 1976 Publications by Franz Moewus on the genetics of algae 310–332. In *The Genetics of Algae* Ed Lewin RA. Oxford, Blackwell Scientific Publications.

Gowers E, revised Greenbaum S, Whitcut J 1986 *The Complete Plain Words* 3rd Edn. London, Her Majesty's Stationery Office. 288 pp.

Gratzer W 2000 *The Undergrowth of Science. Delusion, Self-deception, and Human Frailty*. Oxford, Oxford University Press. 328 pp.

Grime JP, Thompson K, Hunt R and 32 others 1997 Integrated screening validates primary axes of specialisation. *Oikos* **79**: 259–281

Halberg F (undated) *Bibliography 1946–1985*. Minneapolis, Minnesota, Lyon Laboratories. 75 pp.

Hamming R 1986 *You and Your Research:* Lecture at Bell Research Labs Colloquium. <www.cs.virginia.edu/~robins/YouAndYourResearch.html> (6 Sept. 2013)

Hirsch JE 2005 An index to quantify an individual's scientific research output. *Proceedings of the National Academy of Sciences of the USA* **102**: 16569–16572

Hoffmann R 1988 Under the surface of the chemical article. *Angewandte Chemie* **27**: 1593–1602

Hong M-K, Bero LA 2002 How the tobacco industry responded to an influential study of the health effects of secondhand smoke. *British Medical Journal* **325**: 1413–1416

Hubel DH 2009 The way biomedical research is organised has dramatically changed over the past half-century: are the changes for the better? *Neuron* **64**: 161–163. *doi*:10:1016/j.neuron.2009.09.022

Hunt R 1991 Trying an authorship index. *Nature, London* **352**: 187

Ioannidis JPA 2005 Why most published research findings are false. *PLoS Medicine* **2**(8) e124. *doi*:10.1371/Journal.pmed.0020124

IPCC (International Panel on Climate Change) 2007 *Climate Change 2007: The Physical Science Basis*. Cambridge, Cambridge University Press. 996 pp.

Jacks GV 1961 The summary. *Soils and Fertilizers* **24**: 409–410

Johns A 2003 The ambivalence of authorship in early modern Natural Philosophy. In *Scientific Authorship* 67–90. Eds Biagioli M, Galison P. London, Routledge.

Jones RV 1978 *Most Secret War. British Scientific Intelligence 1939–1945*. London, Hamish Hamilton. 556 pp.

Kennedy D 2002 Editorial: Next steps in the Schön affair. *Science* **298**: 495.

Kirkman J 1980 *Good Style for Scientific and Engineering Writing*. London, Pitman. 131 pp.

Knell R 2013 *Introductory R: A Beginner's Guide to Data Visualisation and Analysis in R*. Privately published by R. Knell. Available in e-book format.

Larivière V, Archambault É, Gingras Y, Wallace M 2008 The fall of uncitedness. *Book of Abstracts of the 10th International Conference on Science and Technology Indicators* 279–282.

Lawrence PA 2007 The mismeasurement of science. *Current Biology* **17**(15): R583–R585. *doi*: 10.1016/j.cub.2007.06.014

Läärä E 2009 Statistics: reasoning on uncertainty, and the insignificance of testing null. *Annales Zoologici Fennici* **46**: 138–157

Lock S 1996 Fraud and the editor. In *Fraud and Misconduct in Medical Research* 2nd Edn 240–256. Eds Lock S, Wells F. London, BMA Publishing.

Lock S, Wells F (eds) 1996 *Fraud and Misconduct in Medical Research* 2nd Edn. London, BMA Publishing. 293 pp.

Lozano GA, Lariviere V, Gingras Y 2012 The weakening relationship between the Impact Factor and papers' citations in the digital age. *ArXiv*:1205.4328

Lundh A, Hróbjartsson A, Gøtzsche PC 2012 Income from reprints creates a conflict of interests. *British Medical Journal* **245**: 25.

Maddox B 2002 *Rosalind Franklin. The Dark Lady of DNA*. London, HarperCollins. 380 pp.

Malmer N 1997 Editorial. *Oikos* **77**: 377

Margolis HS, Barwood GP, Huang G, Klein HA, Lea SN, Szymaniec K, Gill P 2004 Hertz-level measurement of the optical clock frequency in a single $^{88}Sr^+$ ion. *Science* **306**: 1355–1358

Massey BS 1971 *Units, Dimensional Analysis, and Physical Similarity*. London, Van Nostrand Reinhold. 140 pp.

Mauas PJD, Flamenco E, Buccino AP 2008 Solar forcing of the stream flow of a continental scale South American river. *Physical Review Letters* **101**: 168501 1–3. *doi*:10.1103/PhysRevLett.101.168501

McCarthy MA 2007 *Bayesian Methods for Ecology*. Cambridge, Cambridge University Press. 295 pp.

McGlashan ML 1971 *Physico-Chemical Quantities and Units*. Monographs for Teachers No. 15. London, Royal Institute of Chemistry. 117 pp.

Medawar PB 1963 Is the scientific paper a fraud? *The Listener* 70: 377–378

Medawar PB 1969 *Induction and Intuition in Scientific Thought. I. The Problem Stated*. Jayne Lectures for 1968. London, Methuen. 76 pp.

Motulsky H, Christopoulos A 2004 *Fitting Models to Biological Data Using Linear and Nonlinear Regression*. Oxford, Oxford University Press. 351 pp.

Moore A 2010 WWDD (What would Darwin do?). *Journal of Evolutionary Biology* 23: 1–5

Noorden R Van 2013 Brazilian citation scheme outed. *Nature, London* 500: 510–511. doi:10.1038/500510a

Orwell G 1946 Politics and the English Language. *Horizon* 13: 252–265

Park R 2000 *Voodoo Science: The Road from Foolishness to Fraud*. Oxford, Oxford University Press. 230 pp.

Peat J, Elliott E, Baur L, Keena V 2002 *Scientific Writing: Easy When You Know How*. London, BMJ Publishing. 292 pp.

Pinker S 1994 *The Language Instinct: The New Science of Language and Mind*. London, Penguin Books. 494 pp.

Press WH, Flannery BP, Teukolsky SA, Vetterling WT 1989 *Numerical Recipes in PASCAL. The Art of Scientific Computing*. Cambridge, Cambridge University Press. 759 pp. [Published in FORTRAN, C, and C++ flavours too.]

Reich ES 2009 The rise and fall of a physics fraudster. *Physics World* 22: 24–29

Rennie D 1999 Misconduct and journal peer review. In *Peer Review in Health Sciences* 90–99. Eds Godlee F, Jefferson T. London, BMJ Books.

Ritter RM 2002 *The Oxford Guide to Style*. Oxford, Oxford University Press. 623 pp.

Roget PM, revised Dutch RA 1962 *Thesaurus*. London, Longman. 1309 pp.

Ross JS, Hill KP, Egilman DS, Krumholz HM 2008 Guest authorship and ghostwriting in publications related to rofecoxib. *Journal of the American Medical Association* 299:1800–1812

Rossner M, Yamada KM 2004 What's in a picture? The temptation of image manipulation. *Journal of Cell Biology* 166: 11–15

Seglen PO 1997 Why the impact factor of journals should not be used for evaluating research. *British Medical Journal* 314: 498–502. doi: 10.1136/bmj.314.7079.497

Smith GCS, Pell JP 2003 Parachute use to prevent death and major trauma related to gravitational challenge: systematic review of randomised controlled trials. *British Medical Journal* **327**: 1459–1461

Sokal A, Bricmont J 1998 *Fashionable Nonsense. Postmodern Intellectuals = Abuse of Science.* New York, Picador. 300 pp.

Sondheimer E, Rogerson A 1981 *Numbers and Infinity.* Cambridge, Cambridge University Press. 172 pp.

Sprent P 1969 *Models in Regression and Related Topics.* London, Methuen. 173 pp.

Strunk W Jr, White EB 2000 *The Elements of Style* 4th Edn. Boston, Massachusetts, Allyn and Bacon. 105 pp.

Student A 1908 The probable error of a mean. *Biometrika* **6**: 1–25

Style Manual Committee, Council of Science Editors 2006 *Scientific Style and Format: The CSE Manual for Authors, Editors, and Publishers* 7th Edn. Reston, Virginia, CSE. 658 pp.

Topol E and 975 others 1993 An international randomized trial comparing four thrombolytic strategies for acute myocardial infarction. *New England Journal of Medicine* **329**: 673–682

Truss L 2003. *Eats, Shoots & Leaves.* London, Profile Books. 209 pp.

Tufte ER 1986 *The Visual Display of Quantitative Information.* Cheshire, Connecticut, Graphics Press. 197 pp.

Tufte ER 1990 *Envisioning Information.* Cheshire, Connecticut, Graphics Press. 126 pp.

Velleman PF, Hoaglin DC 1981 *Applications, Basics and Computing of Exploratory Data Analysis.* Boston, Massachusetts, Duxbury Press. 345 pp.

Venables WN, Smith DM, the R Development Core Team 2009 *An Introduction to R* 2nd Edn. (no place of publication), Network Theory. 137 pp.

Waterston RH and 221 others 2002 Initial sequencing and comparative analysis of the mouse genome. *Nature, London* **420**: 520–562

Wallace ML, Larivière V, Gingras Y 2009 Modelling a century of citation distributions. *Journal of Informatics* **3**: 269–303

Watson JD, Crick FHC 1953 A structure for deoxyribonucleic acid. *Nature, London* **171**: 964–967.

Watson JD 1968 *The Double Helix.* London, Weidenfeld and Nicolson. 226 pp.

Webb T 2008 A bluffers guide to bibliometrics – or, how should we measure science. *Bulletin of the British Ecological Society* **39**(3): 11–14

Webster R 1997 Regression and functional relations. *European Journal of Soil Science* **48**: 557–566

Webster R 2001 Statistics to support soil research and their presentation. *European Journal of Soil Science* **52**: 331-340

Webster R 2002 Analysis of variance, inference, multiple comparisons and sampling effects in soil research. *European Journal of Soil Research* **58**: 74-82

Webster R 2003 Let's re-write the scientific paper. *European Journal of Soil Science* **54**: 215-218

Wieser-Jeunesse C, Matt D, De Cian A 1998 Directed positioning of organometallic fragments inside a calix(4)arene cavity. *Angewandte Chemie International Edition English* **37**: 2861-2864

Worlock K 2013 <www.nature.com/nature/focus/accessdebate/34.html> (6 Sept. 2013)

Young PA 1949 Personal error in penicillin assay. *British Journal of Pharmacology and Chemotherapy* **4**: 366-372

Index

Page numbers in **_bold italic_** are for **tables**; in _italic_ for _figures_

abbreviations, 9, 12
   journal titles, 312, _312_
   Latin, 145
   units, 199
article, scientific, 3–47
   abstract, 10–11
   acknowledg(e)ments, 23
   choosing a target. _See_ journal: choosing a target
   concepts, misunderstood. _See_ words, technical
   conclusions, 23
   discussion, 21–22
   elements in, 13
   how long to publish?, 37
   introduction, 12
   keywords, 9, 10
   letter to editor, submission, 46
   methods, 16–18
   misused words. _See_ words, often misused
   preparation, advice, 39–47
   publication
      duplicate, 37, 110
      fragmented, 35
      salami-slicing, 35
      SPU (smallest publishable unit), 35
   purpose of, 8
   references, 24
   results, 18–21
   'scientific English', 6
   sections, _7_
   solicited, 36
   structure, special
      CONSORT, MOOSE, QUORUM, STARD, 5
   structure, standard, 4
   summary, 10–11
   supplementary materials, 26
   symbols and abbreviations, 12
   symposium or conference, 36
   tables and figures, 25
   title, 7–10
author, scientific, 101–115
   concept of, 107
   criteria for, 113–114, **_114_**
   deluded, 111
   fraudulent, 110

author, scientific (cont.)
  ghost, 112
  graduate and supervisor, 108
  guest. *See* author, scientific: parasitic
  numbers per article, 108
  parasitic, 109

Bayes, 258
Bibliography (for this book), 317

CC (citation count), 28
CI (citation index), 28
CI (confidence interval), 229, 255, 289
citation type
  author-date (Harvard), 306
  citation-number, 307
citing. *See* referencing and citing
CLT (central limit theorem), 226
CMYK. *See* figures: colour
common sense, place for, 260
copyright, 58–62
correlation, 250
  example, egg size and numbers, *256*
  example, river flow and sunspots, *254*
  patterns of, *250*
count, citation. *See* CC
CV (coefficient of variation), 224

data
  types of, *262*
digits, significant. *See* sigdig
dimensions, 200–210
  balance, in equations, 200
  inferring, 210
  logarithms, and, 203
  other than [M], [L], [T], 209
  polynomial coefficients of, 208
  problem solving, *205*
  unbalanced, 202
  units, combining with, 201

distribution
  normal, 223
  Student's '*t*', 227

editor
  duties of, 53
  functions in publishing, 47–53
equations
  dimensions unbalanced, 202
  explicit, 200
  implicit, 202
  implicit, solving, 203
errors
  bias, 214
  bias with imprecision, *215*
  bias, assessing, 216–217
  combination of, 237–241, *238*
    examples, 240, *241*
  imprecision, 214
    normiles, 225
    quantiles, 225
    quartiles, 225
    SD. *See* SD
    SEM. *See* SEM
  imprecision with bias, *215*
  imprecision, quantifying, 221–237
  managing, 211–241
  material, in, 218–220
  measurement, in, 218–220
  mistakes, 212
  types of, *211*, 212
evidence, the. *See* tables and figures

factor, impact. *See* IF
fallacy, planning, 37
figures
  colour in, 277, 297–299
    colour-blind, for the, 299
  design of, 275
  diagrams, line, 294

graphs in. *See* graphs
images, 294–296
modes, colour, 297–299
file, format of, 300–304
   '.asc', '.txt', 300
   '.doc', '.docx', '.odt', '.rtf', 300
   '.pdf', '.ps', '.tex', 300
   '.raw', '.gif', '.png', '.tif', '.jpg', 302
   bitmap. *See* file, format of: rastered
   document layout, 300
   rastered (bitmap), 301–302
   resolution, 301
   text, 300
   vectored, 301
   word-processor, 300
fonts. *See* style: typefaces and fonts
functional analysis, 253, 256
   example, river flow and sunspots, *255*

graphs, 273–294
   bar plots
      categoric, 290
      compared, *291*
      continuous, 292
      discrete, 292
   boxed?, 279
   design of, 275
   misleading, *229*, *256*, *284*
   multiple, arrangement of, 278
   polynomials, dangers of, *287*, 287
   Tufte, 'chart junk', 275
   types of, 290
   'whiskers' about the mean, *229*
graphs, annotations, 289
graphs, axes
   ends (ragged/sharp), 279
   miscellaneous, 284
   orientation, 278
   quotients and ratios, 283

suppressed zero, *277*
ticks, 280
titles, 280
transform, log, 281
zero suppressed, 279
graphs, data
   smoothing, *289*
   symbols, choice of, 286
graphs, lines, fitted, 286
   LOESS, LOWESS, *288*, 289
   regression. *See* regression
graphs, radial
   pie charts, defects of, 293
   wind rose, multivariate data, 293

hypothesis
   testing, 246–250, *251*

IF (impact factor), 29–32, 53
   defects of, 29
   maximum, by subject, *31*
   misuse of, 32
image
   manipulation, unacceptable, 294
importance. *See* significance
index
   citation, 28
   Google Scholar, 32
   Hirsch, 29

jargon. *See* style: jargon
journal
   articles, number of, subject comparison, *103*
   choosing a target, 6, 28–37
   growth
      in articles, *103*
      in authors, *104*
      in pages, *102*
   types of, *33*

330 INDEX

kurtosis, 221

labels, graph axes and table headings, for. *See* variable: name and value
likelihood, 258

mean, 221
median, 221, 223, 225
   symbol for, 221

names, Latin, of organisms, 26–27
NHST (null hypothesis significance test), 257–258
normiles, 225
number, numeral, digit, value, figure, 189

offprints, 46, 57
   as pdf, 46, 57

patents, 58–59
pdf
   for slide design, 76
   offprint, in place of, 46
   proofs, 52, 54
poster, scientific
   at the meeting, 98
   design, *90*
   design, layout, 92–97
   handout. *See* poster, scientific: summary
   leaving for the meeting, 97
   preparing, 89–92
   purpose, 88
   sizes, paper and font, 92, *95*
   summary, 88, 98
   title, 88
proofs, 54–58
   correction marks, *56*
property, intellectual. *See* copyright

publishing
   access, gold or green, 66
   open access, 65
   the digital future, 62
punctuation. *See* style: punctuation

quantiles, 225
quartiles, 225

referee. *See* reviewer
References (for this book), 318–325
referencing and citing, 305–314
   titles, journal, abbreviated, 312, *313*
regression, 251–253
   example, river flow and sunspots, *255*
reprints. *See* offprints
reviewer, 47–50
RGB. *See* figures: colour

science
   processes, types of, 13
SD (standard deviation), 221–225
   change with sample size, *228*
   uses of, 221
SEM (standard error of the mean), 226–230
   change with sample size, *228*
   uses of, 226
SI (Système International), 193–200
   multipliers, 197
   units, *194*
   units, derived, *195*
   units, non-standard, *196*
sigdig (significant digits), 230–237
   examples, *235*
   random values, sampling from, *233*
   rules for, *235*
significance
   is not importance, 260
   P values, 243

skewness, 221, 225
speaking about your work. *See* talk, giving a
standard deviation. *See* SD
statistics
  AIC, 258
  Bayes, *260*
  Bayesian, 258
  'fishing' (mass significance), *246*
  likelihood, 258
  sampling, SD, SEM, *222*
structure, of this book, xiv
style
  abbreviations. *See* abbreviations
  active/passive, 126, 128
  adjectives, use of, 129, 131–132
  ambiguity, 132
  apostrophe, 139
  characters, special, 152
  dates and time, 146
  dominance, first and last elements of, 124
  enclosures, (), [], 142
  footnotes and endnotes, 143
  guidelines, 128–135
  hyphens and dashes, 140
  I/we, 126, 129
  italics, use of, 145
  jargon, 83, 89, 130, 166
  prefixes to names, 144
  punctuation, 136–142
  punctuation, guides, 137
  rules, grammatical, three kinds, 127
  scientific, 125
  typefaces and fonts, 147–152, *152*
    characters, non-keyboard, *156*
  verbs, tense of, 127
  words, misused. *See* words, often misused

words, technical, misunderstood. *See* words, technical
symbols
  chemical, 193
  variables, variates, parameters, 192

tables, 263–273
  author, responsibility for design, 265
  data in, 271
  data, missing, special entries, 272
  entries, order, row/column in, 267
  examples
    data, types of, *262*
    figures, compared with, *273*
    Mendel's peas, *264*
    redesign, power of, *270*
    solar system, *269*
  figures, compared with, *273*
  furniture, 270
  furniture, minimise, 271
  heading/caption, 265
  improvement, by redesign, 268
  layout, 266
  orientation, physical, 267
  portrait/landscape, 267
  row, column, orientation, 267
  strengths, 263
tables and figures
  the evidence, 261–297
talk, giving a, 69–86
  before the performance, 81
  how not to perform, 86
  importance of (Hamming's view), 69
  practice and refine, 80
  purpose, 70
  slide design, *75*
  slide order, 74
  structure of, 72
  summary on paper, 70

talk, giving a (cont.)
   the performance itself, 82
   three performance rules, 82, 85
   topic, choosing a suitable, 70
   translation, perils of, 72
   who will your audience be?, 70
theorem, central limit. *See* CLT
transform, axis
   logarithm, 281

units
   derived. *See* SI: units, derived
   multipliers, SI. *See* SI: multipliers
   non-standard. *See* SI: units, non-standard
   SI. *See* SI: units
   value, part of, 189

value
   in text, digits or words?, 190
   numerical, range of, 190
   thousands separator, 191
variable
   ephemeral, 179
   name and value, 206
      on graph axes, 206, 280
      parenthesis convention, 207
      slash convention, 206
      table rows and columns, 206
   units, logarithm of, invalid, 203
variables, relations among
   correlation. *See* correlation
   functional. *See* functional analysis
   regression. *See* regression
variance, 224

word-processor, 39
   LaTeX, limited acceptability, 39
   mathematical expressions, 39
   pdf, reading, 300
words, often misused (alphabetical), 157
   'affect' to 'frequent', 158
   'gender' to 'may/might', 163
   'maximum' to 'shall/will', 167
   'significant' to 'yes/absolutely', 172
   singular/plural, 172
words, technical, often misunderstood
   abscissa/ordinate, 177
   amount/concentration, 177
   anaerobic/anoxic, 178
   Avogadro constant/number, 178
   citations/references, 178
   constant, 178
   control, 179
   dilute, 179
   efficiency, 179
   ephemeral variable, 179
   flux, 180
   Kelvin/Celsius/centigrade, 180
   light, measurement of, 182
   magnitude, order of, 185
   mol/equivalent, 184
   nutrient/limiting factor, 184
   order, first, second, 180
   parameter/variable, 185
   percent/proportion, 186
   rate, 186
   ratio/quotient, 187
   stress/strain, 187
   variable/variate, 188
   weight/mass/mol, 183